Palgrave Frontiers in Philosophy of Religion

Series Editors
Yujin Nagasawa
Department of Philosophy
University of Birmingham
Birmingham, UK

Erik J. Wielenberg
Department of Philosophy
DePauw University
Greencastle, IN, USA

Palgrave Frontiers in Philosophy of Religion is a long overdue series which will provide a unique platform for the advancement of research in this area. Each book in the series aims to progress a debate in the philosophy of religion by (i) offering a novel argument to establish a strikingly original thesis, or (ii) approaching an ongoing dispute from a radically new point of view. Each title in the series contributes to this aim by utilising recent developments in empirical sciences or cutting-edge research in foundational areas of philosophy (such as metaphysics, epistemology and ethics).

More information about this series at
http://www.palgrave.com/gp/series/14700

E. V. R. Kojonen

The Compatibility of Evolution and Design

palgrave
macmillan

E. V. R. Kojonen
University of Helsinki
Faculty of Theology
Helsinki, Finland

ISSN 2634-6176 ISSN 2634-6184 (electronic)
Palgrave Frontiers in Philosophy of Religion
ISBN 978-3-030-69682-5 ISBN 978-3-030-69683-2 (eBook)
https://doi.org/10.1007/978-3-030-69683-2

This Palgrave Macmillan imprint is published by the registered company Springer Nature Switzerland AG.
The registered company address is: Gewerbestrasse 11, 6330 Cham, Switzerland

Acknowledgments

Issues at the intersection of theology and the natural sciences can be difficult to write about, since it is not easy for one author to attain expertise in all of the various fields involved. This requires especially careful work when one is challenging popular views on foundational matters like the relationship between evolution and divine design. Accordingly, I am indebted to the many rigorous scholars whose work I synthesize and build upon, and to members of the international multidisciplinary community in the field of science and religion for supporting me in this project. Having already explored the debate over design arguments and the relationship between science and religion in previous publications, including *The Intelligent Design Debate and the Temptation of Scientism* (2016), I begun the writing of this book began as part of the "Science-Engaged Theology" project (2017–2019), led by John Perry and Sarah Lane Ritchie at the University of St Andrews. The project provided an ideal enviroment for grappling with the primary scientific literature as part of an international team. I then completed the manuscript during the "God and the Book of Nature" project (2019–2022), led by Mark Harris and Sarah Lane Ritchie at the University of Edinburgh while studying the issue of scientific epistemic authority and its limits. Both projects, and consequently also this publication, were made possible through the support of generous grants from the John Templeton Foundation. At Palgrave, series editor Yujin Nagasawa, Brendan George, and others have also been very supportive throughout the publication process. I am grateful for the opportunity to pursue this research and to have the results published in this series. Naturally, the opinions expressed in this publication are those of the author and do not necessarily reflect the views of the John Templeton Foundation or any of the persons thanked.

Many colleagues have devoted time to discussing with me topics handled in this book, and this has helped sharpen the arguments and consider additional objections. Out of a large group, I wish to thank particularly Zachary Ardern, Andrew Davidson, David Glass, Mikael Leidenhag, Alister McGrath, Jeroen de Ridder, Bethany Sollereder, Christopher Southgate, Aku Visala, Olli-Pekka Vainio, and Mats Wahlberg for their detailed feedback on parts of the manuscript. I also received valuable feedback from the anonymous peer reviewers of the manuscript, members of the Areiopagi project and participants at Helsinki Analytic Theology Workshops. Mark Shackleton from my home University provided feedback on my use of the English language, resulting in many improvements in readability. However, he did not read the final version, so any remaining errors are my own responsibility. Many thanks to everyone, including those I have neglected to mention, for your valuable help and encouragement along the way! And last but not least, I also want to thank my family, particularly my mother Arja, my wife Senja, as well as my children Sofia, Mikael, Aaron, and Nooa for constantly reminding me of the wonder of life. My hope is that something of this wonder can also be glimpsed from the pages that follow.

July 20, 2021 in Järvenpää, Finland — Erkki Vesa Rope Kojonen

Praise for *The Compatibility of Evolution and Design*

"Rope Kojonen offer his readers an informed, nuanced, scholarly, cool, philosophically and scientifically current reflection on the question of what the concepts of 'design' and 'purpose' might mean within evolutionary theory. His approach is necessarily disarming, subtle and gently corrosive to all entrenched positions. No established point of view escapes unchallenged—as an example of how philosophical and theological thinking can spur imaginative progress in scientific reasoning, it is a welcome and important contribution."

—Tom McLeish, *Professor of Natural Philosophy in the Department of Physics, University of York, UK*

"It shows considerable courage to re-invoke the term 'design' in relation to the theology of biology. The concept seemed tellingly dismissed by Richard Dawkins, and revived only in highly dubious form by intelligent design theorists. So Rope Kojonen deserves great credit for re-evaluating the idea of divine design in dialogue with the whole range of contemporary literature. This book will fill a gap in that literature, and should be of considerable help to teachers and students alike."

—Christopher Southgate, *Professor of Christian Theodicy, University of Exeter, UK*

"A remarkably helpful and important intervention in the long-standing debates about arguments from design, natural theology, and the impact of evolutionary theory on Christian theology and apologetics. Kojonen's work enables a richer and fuller engagement with these important topics, while opening up fresh lines of discussion and exploration, especially in relation to how we should understand—and might also discern—design within the natural world."

—Alister E. McGrath, *Andreas Idreos Professor of Science and Religion, University of Oxford, UK*

"This book offers a well-researched, lucid, carefully argued, and well-balanced assessment of whether evolutionary theory is compatible with holding that biological teleology provides evidence of a Creator. It is an essential reading for anyone interested in the relationship between science, philosophy, and religion."

—Andrew T. Loke, Associate Professor, *Hong Kong Baptist University, Hong Kong*

Contents

1 **Introduction: Paley's Cosmic Temple and the Darwinian Critique** 1
1.1 Nature as a Temple 1
1.2 The Task of This Book 4
1.3 The Structure of the Book 8
References 11

2 **Reclaiming the Biological Design Argument: A Preposterous and Dangerous Idea?** 13
2.1 Between Scientism and Fideism 14
The Philosophy of Scientism 14
Faith and Reason 17
God the Designer? 22
2.2 On the Compatibility of Evolution and Creation 26
The Assumption of Conflict 26
Divine Action and Chance 28
Beyond Mere Compatibility 30
2.3 Perceiving Design in Nature 32
Design Arguments and the "Theist on the Street" 32
Design Discourse Rather Than Design Arguments? 34
Debunking Arguments 39
2.4 Conclusions 44
References 47

3 From Teleological Arguments to Thought Experiments 59
3.1 Explaining Evidence of Design 61
Personal Explanation 61
The Fine-Tuning Design Argument 65
Biological Design 69
St Thomas vs. Paley and ID 71
3.2 The Logic of Design Arguments 75
Analogical, Deductive, or an Inference to the Best (or Only) Explanation? 75
Thought Experiments and Objections 79
3.3 Conclusions 86
References 90

4 Not by Selection Alone: Evolutionary Explanations and Their Requirements 97
4.1 The Darwinian Blind (but Powerful) Watchmaker 98
The Basics of Evolutionary Explanations 98
Explaining Complex Adaptations 105
4.2 Answering Critiques of Evolutionary Explanations 109
The Conservation of Information 109
Irreducible Complexity 115
Protein Evolution 119
4.3 Teleology All the Way Down? 123
Laws of Form and Convergence 123
Determinism and Indeterminism 128
4.4 Conclusions 132
References 135

5 Indirect Design and the Revelatory Potential of Biology 145
5.1 Design and Natural Explanations 146
Compatibility or Incompatibility? 146
Developing an Account of Compatibility 149
5.2 From Possibility to Plausibility 157
Defending Design Discourse 157
Analogies of Indirect Design 164

5.3 *Dealing with the Problem of Natural Evil* 174
Understanding the Problem 175
Evaluating Evolutionary Theodicies 181
The Freedom of Evolution and the Responsibility of the Creator 187
5.4 *Conclusions* 193
References 198

6 Conclusion: From Teleology to Theology, and Back Again 205
References 213

Index 215

CHAPTER 1

Introduction: Paley's Cosmic Temple and the Darwinian Critique

1.1 Nature as a Temple

In his famous book *Natural Theology* (2008 [1802]), William Paley argues that in light of the evidence of design that biological organisms exhibit, the world can be compared to a great temple, reflecting the Creator's glory:

> The world from thenceforth becomes a temple, and life itself one continued act of adoration. The change is no less than this, that whereas formerly God was seldom in our thoughts, we can now scarcely look upon anything without perceiving its relation to him. Every organized natural body, in the provisions which it contains for its sustentation and propagation, testifies a care on the part of the Creator expressly directed to these purposes.[1]

Paley begins his book with the famous example of a watch found on a heath. Suppose that we have no prior knowledge of watches, but want to understand how they originated. According to Paley, the "contrivance" of the watch, the way its parts fit together just in the way that allows its mechanism to function, would immediately lead us to reason that it must have a design, rather than believing that the watch appeared by chance. Applying this analogy to nature, Paley argues that biological organisms possess far superior contrivance, and thus should also lead us to the conclusion of design.

E. V. R. Kojonen, *The Compatibility of Evolution and Design*, Palgrave Frontiers in Philosophy of Religion,
https://doi.org/10.1007/978-3-030-69683-2_1

Paley's *Natural Theology* was familiar to Charles Darwin, and Darwin concurred with the idea that the apparent contrivance of biological organisms does need an explanation. However, Darwin would later come to reject Paley's watchmaker analogy. In a famous passage in his autobiography, Darwin (1958 [1887], 87) wrote that "the old argument of design in nature, as given by Paley, which formerly seemed to me so conclusive, falls, now that the law of natural selection has been discovered. We can no longer argue that, for instance, the beautiful hinge of a bivalve shell must have been made by an intelligent being, like the hinge of a door by a man." Darwin acknowledged the real need to explain why the order of the eye appears to be directed for the purpose of enabling vision, and the order of the heart is so suited for pumping blood. It does not seem credible that such orderliness would have come about simply by chance. However, Darwin's insight was that these could be explained as the result of evolutionary mechanisms, without needing to appeal to the Creator in any part of the explanation, at least in any sense different from explaining other natural phenomena like the "course which the wind blows" (Darwin 1958 [1887], 88). Biology was now secularized and naturalized (Walsh 2008; Ruse 2003).

Michael Ruse thus argues that Darwinism fundamentally altered the dominant religious interpretation of nature. Whereas before Darwin, nature sang of its Creator, "in the post-Darwinian World, not only does nature not sing, it is hard to hear the tunes of heaven either. [...] After Darwin, meaning has been drained from the world and all is laid on faith" (Ruse 2018, 282). Thus, while Ruse argues that evolutionary theory is compatible with religious faith, he nevertheless thinks that evolution had an impact on the plausibility of the idea of design. Anti-evolutionists also commonly see this as a point of tension. In my previous study on the debate surrounding the Intelligent Design (ID) movement (Kojonen 2016a), I noted that the movement's thinkers see the question of evidence of design in biology as crucial for assessing the religious implications of evolutionary theory. William A. Dembski, one of the central intellectual architects of the movement, writes that

> Within theistic evolution, God is a master of stealth who constantly eludes our best efforts to detect him empirically. Yes, the theistic evolutionist believes that the universe is designed. Yet insofar as there is design in the universe, it is design we recognize strictly through the eyes of faith. Accordingly the physical world in itself provides no evidence that life is

designed. For all we can tell, our appearance on planet earth is an accident. (Dembski 1999, 110)

Elsewhere I have argued that the ID movement's critique of theistic evolutionism is ultimately in conflict with the way core ID proponents understand the relationship between religious belief and science (Kojonen 2013, 2016a, 169–189). However, here the central point is that Ruse and Dembski present a similar historical narrative: before the Darwinian revolution, one could think that the order of biological nature sings of the Creator. However, insofar as one accepts evolution as a good explanation of biology, then theistic evolution—the belief that this process was guided by the Creator—is possible only through a leap of faith. On the popular level, it is even more common to argue that Darwinian theories challenge the credibility of Christian belief in creation.

This conclusion seems to go too far, since, as I will argue in the next chapter, the rationality of religious belief does not require biological design arguments. Nevertheless, the common view even among theistic evolutionists is also that evolutionary theory does indeed make biological design arguments obsolete. Insofar as theistic evolutionists use arguments in support of the rationality of belief, these are based on other features of the cosmos, such as what Alister McGrath (2009) calls the "wider teleology" of the broader cosmos that makes evolution possible.

Many theistic evolutionists even see the demise of the biological design argument as a blessing. For example, Michael Hanby (2016, 165) argues that Paley's design argument rests on a shallow theology, and cannot attain to knowledge of God for "the simple reason" that Paley's "contriver is not God." Conor Cunningham (2010) argues that evolution helps theology by purifying it from faulty God of the gaps -type reasoning, and forcing theology to the classical understanding of God as the ground of all of existence.[2] Francisco Ayala (2007) even argues that design arguments come close to blasphemy by making God responsible for all the poorly designed features of organisms. Thus, rather than representing the zenith of Christian natural theology, as in Ruse's account, it has been argued that Paley's thought actually already compromised core Christian ideas, and ID's design arguments represent a continuation of this compromise.

In this book, I will not defend the correctness of Paley's or ID's biological design arguments as such. The ID movement and its Darwinian critics understand evolution and design as mutually exclusive explanations, functioning on the same explanatory level, as answers to the same questions.[3] If understood in this way, biological design arguments are indeed in tension (or even in contradiction) with evolutionary biology. Furthermore, as Darwin states, it seems clear to me that evolutionists cannot see organisms as designed in the same way as human craftsmen make door hinges. To be fair, despite the criticism by Hanby and others, it is not clear to me that Paley's or the ID movement's design arguments make such a denial of divine transcendence necessary (Kojonen 2016a, 89–96). Insofar as the design argument is in practice accompanied by an overly anthropomorphic understanding of God, then in leading to a rejection of it, Darwinism would indeed force believers to steer away from such a faulty understanding. But is this really the only way to understand the design argument?

1.2 The Task of This Book

This assumption of an explanatory competition between evolution and design has had enormous influence on discussions of creation and evolution both on the academic and the popular level. However, after extensive reading and deliberation, it now seems to me that this assumption is not nearly as well-founded as is often claimed. Using the methods of philosophical analysis, my purpose in this book is to bring to light the reasons behind this assumption, and then to develop an alternative view. Taking the evolutionary objection seriously, I will nevertheless argue in this book that the teleological order of biological organisms can still, in a rationally permissible way, be understood as a sign of the divine reality, even in an evolutionary cosmos.

This type of argument is not without historical precedent. Darwin's friend, the American botanist Asa Gray (1810–1888), famously disagreed with Darwin on the relationship of design and evolution. According to Gray (1860), evolution actually "leaves the question of design just where it was before," because the biological design argument does not in any way depend on whether God created living organisms directly, through miracles, or through a secondary cause such as evolution. Seeing the end result, Gray claims, is still enough to create a compelling case for design. In the history of the science and religion discussion, Darwin's own interpretation of the effects of evolution on the biological design argument has largely carried the day.[4]

Nevertheless, in recent discussion, several philosophers and theologians have defended the possibility that evolution indeed does not undermine design beliefs that are based on the order of biology, contrary to the usual view. Many now believe that the scientific data also supports belief in at least some directionality in evolution. It seems to me that now, with new work in the philosophy of biology, the philosophy of religion, and the natural sciences, it is again plausible to understand biological order as the supreme manifestation of the "wider teleology," and thus as also indirectly revelatory of the Creator. The idea that biological nature provides us with evidence of design can thus be reclaimed for use within a theistic evolutionist understanding of nature. This promises to transform the debate over design and evolution, not only providing a defense of the logical compatibility of evolution and creation, but also moving beyond this mere compatibility by rehabilitating the idea of evidence for design.

As the following chapters will show in detail, it seems to me that the easiest way to defend the compatibility of design and evolution is in the context of a Christian theology of nature, interpreting nature based on theological presuppositions. However, in order to be part of a credible understanding of nature, this interpretation of biology cannot be forced, but should be consistent with the evidence. The "reading in" of purpose to nature, in order to be valid, requires that there must be some sense in which the phenomena of teleology in nature indeed make sense from a theistic perspective, and can thus also be "read out of" nature to some extent. Thus, if the salvaging operation is to be successful, the evidence of biological teleology needs to be shown to provide at least some support for the rationality of religious belief, even while accepting evolutionary explanations as true or probably true.

In the nineteenth-century reception of evolutionary theory, Asa Gray was not the only one to defend the compatibility of design arguments and evolutionary biology. In the same vein, theologian Charles Kingsley (1874, xxvii) famously wrote that "we knew of old that God was so wise that He could make all things: but behold, He is so much wiser than that, that He can make all things make themselves." Mats Wahlberg (2012, 182), one of the contemporary defenders of the revelatory potential of biology, comments: "If it takes more wisdom to create through an evolutionary process than by hands-on-design, and if structures created by hand-on-design by humans are expressive of human intent and intelligence, why could not structures created by God in that more wisdom-demanding way reflect divine intent and intelligence?" In the following chapters, I will argue for the correctness of this intuition. The production of biological organisms is a demanding process, requiring highly fine-tuned natural laws and building blocks. If the "wider teleology" of the cosmos that underlies the possibility of biological organisms is indeed revelatory of God's wisdom, as McGrath and other contemporary theistic evolutionists argue, then it stands to reason that the products of this wider teleology are also revelatory of God. The order of biology itself provides grounds for believing in a purposeful Creator, and successful evolutionary explanations do not remove these grounds, although evolution does modify our understanding of design. This means that the basic intuitions underlying biological design arguments are compatible with evolutionary biology. However, the biological design argument developed here should be understood as a primarily philosophical argument, rather than a scientific one—just as the thesis of opposition between design and evolution is also philosophical, not scientific.

In my view, however, the classification of design arguments as part of the broad categories of "science" or "philosophy" is ultimately of little consequence in comparison to the question of whether the arguments are supported by good reasoning and evidence (Kojonen 2017). If we adopt the popular conception of science as methodologically naturalistic, then the question of design should be understood as nonscientific. Nevertheless, the features of the natural world, as studied by the natural sciences, will still be relevant for the argument. Some readers might even question whether philosophical and theological analysis could have anything to contribute to the discussion beyond what the natural sciences have to say. I will address these worries in the coming chapters. Here I will simply state that the effect of philosophical and theological considerations seems unavoidable. Questions like "when does one explanation eliminate another," "could the order of the world even in principle reveal a Creator" and "how does suffering fit with the idea of design" are necessary for the debate and require moving beyond just scientific considerations, to considering the nature of explanation and broader metaphysical story we believe about the world. Moreover, the position that evolution and divine design do not fit together is also a philosophical and even a theological (or antitheological) conclusion, requiring just as much philosophical justification as the conclusion of compatibility. For these and many other reasons, I think addressing the question of design and its compatibility with evolution requires interdisciplinary analysis. And hopefully this also helps justify why I am writing about these things as a philosopher of religion and a theologian, and do not wish to leave it just to biologists and philosophers of biology. My goal has nevertheless been to engage even with the primary scientific literature to the best of my ability, for I think evolutionary biology does deserve to be taken seriously by non-biologists as well.

Theodosius Dobzhansky's (1973) famous statement "nothing in biology makes sense except in the light of evolution" succinctly expresses the fact that evolution is now a unifying theory for all of biology. Evolution is used to explain innumerable details that, according to the vast majority of biologists, would otherwise be left without explanation. In the period of over 150 years since the publication of Darwin's *Origin of Species*, a massive amount of biological research has been conducted from the evolutionary perspective. Dobzhansky and contemporary evolutionary biologists admit that not all details about how evolution happens have been worked out, but they nevertheless argue that this should not stop us from accepting as correct the general picture given by the theory. The logic of the inference to the best explanation does not demand that all questions related to a theory, such as evolution, are resolved before it can be affirmed as a better theory than any of its rivals. Now, it may be that the current mainstream view of evolution will ultimately turn out to be wrong or lacking in some respect. Within evolutionary biology itself, there is a considerable amount of discussion about expanding and modifying the theory, along with the recognition that many puzzles still remain. Proponents of Intelligent Design and other forms of anti-evolutionism see the theory as fundamentally wrong. However, in this book, the validity of the essential scientific claims of evolutionary biology will be accepted as the starting point of my inquiry. I invite proponents of ID, as well, to assume the plausibility of evolution for the sake of the argument, and to join me in asking if the falsity of the conclusion of design really follows from it. Suppose that the evolutionary mechanisms biologists study really are able to generate the wonders of biology from hummingbirds to human brains. Is the production of such results then more plausible if those mechanisms were purposefully designed, or if they were not? Could the products of an indirect evolutionary process even in principle tell us something about the rationality of the cosmos and the wisdom of its Creator?

Talk of a salvaging operation for saving some value out of the biological design argument calls to mind the raising of an old hulk, covered in crustaceans, from the depths of the ocean and either repairing it for a fresh voyage or repurposing its parts. For many on both the popular level and in academia, the idea of biological design has been relegated to the junk heap, but it seems to me that this rejection has been premature. Thus, this book represents my own attempt at re-imagining how biological design as a sign or evidence of the Creator can still be held even by those who accept evolution. As such, this book is a follow-up to my previous study on design arguments, *The Intelligent Design Debate and the Temptation of Scientism* (Kojonen 2016a). In that book, I analyzed the design argument of the ID movement, attempting to understand both the defenses and critiques of that argument. During the research that went into that book, I noticed that the incompatibility of Darwinian science and biological design arguments was commonly assumed by both critics and defenders of the movement. However, I began to suspect that this assumption was not necessarily correct. To my delight, I found that others had made similar observations in the literature, though the argument for the compatibility of design and evolution has gone largely unnoticed in popular-level discussions. In this book, I do my best to lay out the grounds for both the traditional view of incompatibility and the rationality of simultaneously accepting both design and Darwinian explanations as true.

1.3 The Structure of the Book

The book's argument is divided into four main chapters in addition to this introduction and the conclusion. In Chap. 2, I explain the nature of my salvaging operation further, and clarify its relationship to other views in the creation/evolution discussion. The very idea of design arguments is controversial, and has been criticized from various theological and philosophical perspectives. For example, it has been argued that the language of design is too anthropomorphic, and is hence unfit for the classical understanding of God as Creator. I lay out some of the core background assumptions that inform the book, and argue in favor of the theological value of the idea of design as part of our understanding of nature. I also defend the general necessity of a philosophical and theological assessment of this issue, in contrast to scientistic views. The relationship of design arguments to intuitive design perceptions is also an important issue that I deal with in this chapter. While I do argue that we can salvage the design argument in light of evolution, I also argue that it would already be valuable to be able to harmonize intuitive design perceptions (or "design discourse," as discussed by Alvin Plantinga, 2011) and evolution.

Before being able to understand the claim that design arguments and evolution must be opposed, we must first understand how design arguments work in the first place. In the third chapter, I thus analyze the basics of design arguments. There is a debate between those who hold that philosophical considerations are sufficient to undermine the biological design argument, and those who hold that belief in biological design was the most rational position before the Darwinian revolution. Of course, if the philosophical objections were sufficient by themselves, then there would be no need to discuss the topic of evolution. So, to lay the groundwork for the following chapters, I also argue that the Humean objections by themselves are not sufficient as counterarguments, and that the design argument can be constructed in a way that bypasses the philosophical objections.

In the fourth chapter, I analyze the claim of the incompatibility of evolution and design, as well as the debate over the power of the evolutionary mechanism. After all, before trying to solve the proposed contradiction between evolution and design, we should first make sure that we understand what that contradiction is supposed to be. Evolutionary processes are supposed to provide an explanation for precisely the same features of biology that design arguments also attempt to explain—therefore making design unnecessary. This is based both on an understanding of simplicity as a virtue of theories, and on a particular understanding of evolution as producing features due to way natural selection and other evolutionary mechanisms function in contingent historical circumstances. I argue, however, that the scientific data point to a more important role for evolutionary mechanisms other than selection, as well as the dependency of evolution on the "wider teleology" of the cosmos. As biologists continue to posit "laws of form" that provide some directionality of evolution, this has the effect of also weakening the opposition between design arguments and evolutionary explanations.

In the fifth chapter, I argue in more detail for the compatibility of design and evolution as an explanation, examining how the evidence from biology continues to support belief in design. I argue that we can save both the rationality of an intuitive design discourse and the rationality of seeing biological teleology as evidence of design, even when adopting an evolutionist view of biology. Building on the previously discussed empirical evidence as well as on philosophical considerations, I analyze several different ways of harmonizing design and evolution as explanations, such as the level-shifting strategy, understanding evolution and design as answers to different questions, and design discourse. I also contrast my approach to other views, such as Plantinga's design discourse. I also develop possible objections and analyze them through various thought experiments and analogies. One of the most important objections the chapter deals with is the problem of natural evil. This problem was also central for Darwin (1856), who asked "what a book a devil's chaplain might write on the clumsy, wasteful, blundering, low and horridly cruel works of nature," though he also argued that the products of evolution are "endless forms most beautiful" (Darwin 2008 [1859], 360). It has been argued that evolution greatly intensifies the problem of natural evil, and this might be seen as making an evolutionary design argument unviable. It has also been argued that evil should lead us to prefer a theology which does not make God responsible for biological order, which would also undercut the design argument. I argue that a design argument combining evolution and design can respond to all these concerns as well as any other theological approach, and may even help in better responding to the problem.

The topic of the relationship of evolution and religious beliefs is one of the most discussed themes in the field of science and religion, and I feel that making a fruitful contribution to this discussion requires extensive dialogue with the existing literature. The long intellectual history of the opposition between design and evolution requires one to address many different objections even to the whole prospect of salvaging something of value from the biological design argument. To do justice to the vast literature on the topic, the method of analysis used in this book blends elements from both the continental and the analytic approaches. As is usual in the Nordic and continental tradition of systematic theology, I typically begin each chapter by analyzing, contrasting, and developing arguments that already exist in the literature, only then forging my own path in the usual manner of Anglo-American analytic philosophy and theology. This will, I believe, bring to light clearly how my position on the relationship between design and evolution differs from other views, and will help readers to themselves evaluate the reasons given.

Notes

1. Paley (2008 [1802], chapter XXVII). For the historical discussion, see particularly McGrath (2009). The metaphor of the cosmos as a temple can be traced back to Genesis 1, and has been noted since patristic times. Isaiah 66:1-2 similarly argues that creation itself is God's temple. See Brown (2010, chapter 3).
2. The phrase "God of the gaps" goes back at least to Dietrich Bonhoeffer, who wrote that "it is wrong to use God as a stopgap for the incompleteness

of our knowledge. If in fact the frontiers of knowledge are being pushed further and further back (and that is bound to be the case), then God is being pushed back with them, and is therefore continually in retreat. We are to find God in what we know, not in what we do not know; God wants us to realize his presence, not in unsolved problems but in those that are solved" (Bonhoeffer 1997, 311). However, in the present discussion the term has often been used in a very ambiguous way, which I have attempted to clarify elsewhere (Kojonen 2016b).

3. The metaphor of levels of explanation is commonly used, but has also been criticized. Another possibility is to simply discuss different explanations as answers to different questions. For a discussion using the example of neuroscience, see Craver (2007).
4. See particularly Johnson (2015) and Browne (2010).

References

Ayala, Francisco. 2007. *Darwin's Gift to Science and Religion*. Washington, DC: Joseph Henry Press.

Bonhoeffer, Dietrich. 1997. *Letters and Papers from Prison. New Greatly*. Enlarged Ed. New York, NY: Touchstone.

Brown, William P. 2010. *The Seven Pillars of Creation: The Bible, Science and the Ecology of Wonder*. Oxford: Oxford University Press.

Browne, Janet. 2010. Asa Gray and Charles Darwin: Corresponding Naturalists. *Harvard Papers in Botany* 15 (2): 209–220.

Craver, Karl. 2007. *Explaining the Brain: Mechanisms and the Mosaic Unity of Neuroscience*. Oxford: Oxford University Press.

Cunningham, Conor. 2010. *Darwin's Pious Idea: Why the Ultra-Darwinists and Creationists Both Get It Wrong*. Grand Rapids, MI: William B. Eerdmans.

Darwin, Charles. 1856. Letter to J. D. Hooker, dated July 13, 1856. *Letter No. 1924. The Darwin Project*. Available at https://www.darwinproject.ac.uk/letter/DCP-LETT-1924.xml.

———. 1958 [1887]. In *The Autobiography of Charles Darwin*, ed. Nora Barlow. London: Collins.

———. 2008 [1859]. *On the Origin of Species by Means of Natural Selection, Or, the Preservation of Favoured Races in the Struggle for Life. 2nd Edition (1859). Oxford World's Classics*. Oxford: Oxford University Press.

Dembski, William A. 1999. *Intelligent Design: The Bridge between Science and Theology*. Downers Grove, IL: InterVarsity Press.

Dobzhansky, Theodosius. 1973. Nothing in Biology Makes Sense Except in the Light of Evolution. *American Biology Teacher* 35 (3): 125–129.

Gray, Asa. 1860. Natural Selection Not Inconsistent with Natural Theology. In *Atlantic Monthly for July, August and October, 1860. Darwiniana: Essays and Reviews Pertaining to Darwinism*. New York: D. Appleton. 1888 (1874).
Hanby, Michael. 2016. *No God, No Science: Theology, Cosmology, Biology*. Oxford: Wiley-Blackwell.
Johnson, Curtis. 2015. *Darwin's Dice: The Idea of Chance in the Thought of Charles Darwin*. New York, NY and Oxford: Oxford University Press.
Kingsley, Charles. 1874. *The Natural Theology of the Future*. London: Macmillan.
Kojonen, Erkki Vesa Rope. 2013. Tensions in Intelligent Design's Critique of Theistic Evolutionism. *Zygon* 48 (2): 251–273.
———. 2016a. *The Intelligent Design Debate and the Temptation of Scientism*. London: Routledge.
———. 2016b. The God of the Gaps, Natural Theology and Intelligent Design. *The Journal of Analytic Theology* 4. https://doi.org/10.12978/jat.2016-4.041708101413a.
———. 2017. Methodological Naturalism and the Truth Seeking Objection. *International Journal for Philosophy of Religion* 81 (3): 335–355.
McGrath, Alister. 2009. *Darwinism and the Divine: Evolutionary Thought and Natural Theology*. Oxford: Wiley-Blackwell.
Paley, William. 2008 [1802]. *Natural Theology*. Oxford World's Classics. Oxford: Oxford University.
Plantinga, Alvin. 2011. *Where the Conflict Really Lies: Science, Religion, and Naturalism*. New York, NY: Oxford University Press.
Ruse, Michael. 2003. *Darwin and Design: Does Evolution Have a Purpose?* Cambridge, MA: Harvard University Press.
———. 2018. *Darwinism as Religion: What Literature Tells Us About Evolution*, 282. Oxford: Oxford University Press.
Wahlberg, Mats. 2012. *Reshaping Natural Theology: Seeing Nature as Creation*. London: Palgrave Macmillan.
Walsh, Denis. 2008. Teleology. In *The Oxford Handbook of Philosophy of Biology*, ed. Michael Ruse, 113–137. Oxford: Oxford University Press.

CHAPTER 2

Reclaiming the Biological Design Argument: A Preposterous and Dangerous Idea?

In stark contrast to the view that design arguments are theologically valuable, Cunningham (2010, 278) claims that "it seems perfectly fair to say that if ID has gotten what it thinks it wants (inference to a designer), then it would amount to one of the most significant arguments for atheism ever—for the designer, now proved, would not be worthy of worship." In light of the heavy criticism faced by Paley and the Intelligent Design movement, and the common sentiment that evolution and design are incompatible, any attempt at arguing for the compatibility of design and Darwinian explanations is bound to raise questions. The polemical nature of the debate surrounding the ID movement, in particular, makes it necessary for me to be careful to show how my argument relates to other views. For many participants in the debate, it is a battle between good and evil—where it is difficult to see any good ideas on the other side. In this situation, the aim of this chapter is to provide some necessary background for the argument of this book.

First, it is important to clarify the relationship of this argument with the natural sciences. Does attempting to salvage something of value from biological design arguments not constitute a theological attack on the natural sciences, a doomed "God of the gaps" view? This is a common criticism of the biological design argument. Yet if the design argument is not a scientific argument, then others would argue that this, then, also spells doom for the argument, as the limits of science are, for them, also the limits of

E. V. R. Kojonen, *The Compatibility of Evolution and Design*, Palgrave Frontiers in Philosophy of Religion,
https://doi.org/10.1007/978-3-030-69683-2_2

reliable empirical knowledge. I will argue that this dilemma unnecessarily restricts our options.

Second, the idea that we could see design in an evolutionary world is bound to strike some as preposterous, given the enormous influence of the idea of a conflict between evolution and creation on the popular level, as well as concerns about the compatibility of randomness and purpose. In contrast, the present argument builds on the basic compatibility of creation and evolution, as established in the field of science and religion. But I will argue that it is also important to defend the idea of the revelatory potential of creation, going beyond the mere compatibility of the doctrine of evolution and creation, to also consider the compatibility of design and evolution.

Third, the theological value of this salvaging operation needs to be analyzed, in light of the theological critique against design arguments, such as the one Cunningham presents. Supposing that it is relevant to consider evidence for and against religious beliefs, this does not then mean that the design argument actually supports religious belief. For instance, if design arguments compromise God's transcendence, then why should anyone even want to reclaim them for use in our understanding of nature? In the remainder of this chapter, I will respond to each concern, and argue for the value of the idea of design.

2.1 Between Scientism and Fideism

The Philosophy of Scientism

The critique of design arguments as God of the gaps arguments is common in the literature, particularly in popular-level treatments. Dawkins, for example, argues that:

> Admissions of ignorance and temporary mystification are vital to good science. It is therefore unfortunate, to say the least, that the main strategy of creation propagandists is the negative one of seeking out gaps in scientific knowledge and claiming to fill them with 'intelligent design' by default. [...] It is precisely the fact that ID has no evidence of its own, but thrives like a weed in gaps left by scientific knowledge, that sits uneasily with science's need to identify and proclaim the very same gaps as a prelude to researching them. (Dawkins 2006, 152–153)

I have argued elsewhere that this characterization of the movement's arguments is unfair. Proponents of ID argue based on what they think we know about intelligent causes, life, and the limits of natural processes, rather than based merely on our supposed ignorance of plausible natural explanations. However, it is true that they do see explanatory gaps in evolutionary mechanisms as crucial for their design argument. In contrast, the mainstream view is that we have good evidential support for the proposition that current gaps in evolutionary explanations can eventually be filled in without supposing interventions by a designer (Kojonen 2016b).

In contrast to ID's argument, the salvaging operation developed here does not require gaps in the evolutionary processes. Nevertheless, the argument does require the possibility of rational non-naturalistic explanations—and some would see all theistic explanations as "Gods of the gaps," no matter what area they are presented in. For the argument of this book to be possible, the natural explanations explored by science cannot be the only ones that are possible already in principle. Indeed, John F. Haught (2004, 238) argues that the problem with "God of the gaps"-style explanations is that they appeal to God "at a point in inquiry when there is still plenty of room for further scientific elucidation." Putting the point in this way, the critique of the God of the gaps argument does not imply that we could not in principle find limits to scientific inquiry, or other bases for appealing to God as an explanation.

The exact proper limits of scientific and philosophical or theological inquiry are not easy to settle, however. Richard Swinburne (2004a) famously argues that explanations involving God are a type of personal explanation, and are legitimate for explaining phenomena which are "too big" or "too odd" for science, such as the general orderliness of the cosmos. Ian G. Barbour, a pioneer in the field of "religion and science," defines limit questions as "ontological questions raised by the scientific enterprise as a whole but not answered by the methods of science" (Barbour 1997, 90). Using strategies like these, theistic evolutionists can argue that the domain of natural science does not encompass everything, and there is room for non-scientific rational beliefs. Though science may also have a bearing on such big questions or limit questions, it does not itself provide a full answer. Against those who would see the interaction of science and religion as mainly a conflict, theistic evolutionists argue for the possibility of positive interaction and coexistence.[1]

In contrast, others argue for the thesis of scientism, which states that science is our best or even our only reliable "guide to reality" (Rosenberg

2011), and that philosophical and theological metaphysics and explanations are of no value, containing only useless "sophistry and illusion," to use Hume's famous description (Hume 1991 [1748], 87). From this perspective, all explanations involving God are examples of "God of the gaps" reasoning. The idea is that the boundaries of science will continue to expand over time, and if the explanation for some class of facts remains forever outside the reach of science, then surely no other rational explanations of those facts can be made either. In my view, the danger here is that of replacing the "God of the gaps" with a "naturalism of the gaps," in which all unsolved phenomena are simply assumed to have natural explanations, no matter how different these phenomena are from the ones that have so far been explained within the natural sciences. This book is based on the rejection of this kind of scientism—which in my opinion distorts both defenses and critiques of design arguments. For example, the debate concentrates too much on the question of whether design arguments can be properly called science or not (Kojonen 2016a).[2]

Scientism is a philosophical idea that should be distinguished from science itself, as there is no scientific result that could demonstrate that science is the only way to gain knowledge, or that scientific explanations are the only valid ones. Evidence for the successfulness of the natural sciences does nothing to diminish the rationality of a philosophical design argument, for example. Instead, what would be required is a direct critique of the rationality of theology, philosophy, and everyday reasoning. A central problem is that scientism as a philosophical idea appears to be self-defeating: in order to be able to do learn and practice science, we must be able to rely on the basic reliability of our everyday reasoning and perception. Without this basic reliability, the scientist could not attain to beliefs such as "the peer-reviewed literature states X," "my research instruments show Y," or "this experimental result does not support my hypothesis." The scientific process builds on accumulated and intersubjectively tested sensory experience. But "garbage in, garbage out": if ordinary experience did not have even a little validity, then accumulating it would not help either. Moreover, it is not clear that scientific arguments always give more certain knowledge than ordinary experience or philosophical arguments. For instance, the basic existence of conscious experience and the external world seems to be more clearly true than even clearly established scientific truths like the heliocentric model of the solar system.[3]

Such arguments, discussed extensively in the literature, make it credible to think that philosophical and theological reasoning could also lead to justified true beliefs about questions related to the designedness or non-designedness of the cosmos—but the scientific evidence should certainly be properly taken into account. Indeed, in my previous study of the ID debate, I argued that all major positions in the debate derive at least part of their vigor from theological and philosophical factors, not merely science. For example, some of the core arguments against ID, such as the problem of bad design, are essentially theological arguments, while the problem of the God of the gaps is both a philosophical and a theological criticism.

In Chap. 3, I will further analyze design as an explanation, and ask whether design-based explanations might possess explanatory power even in some situations where they do not meet the criteria of science. In Chaps. 4 and 5, I will analyze in more detail what questions about biological teleology might be left unanswered by evolutionary science, and in what sense it might be reasonable to think that these questions are better answered by design, without assuming gaps in evolutionary processes. One example of a limit question might be: "What ultimately explains the possibility of evolving biological teleology?" And if the response to this question is theistic, then can we truly say that the evolutionary process works without God? Or can we credibly claim that God is not in some sense responsible for, and to be credited for, the existence of biological order—both the apparently beneficial teleology of biology and then also (perhaps problematically) the more horrifying parts? And if he is responsible, then can the order of biological organisms provide some evidence for God's existence?

Faith and Reason

One's view of the theological value of the salvaging operation will depend partly on one's overall view of the relationship between faith and reason, and one's understanding of God. The rejection of scientism already implies that religious belief in God does not necessarily need to be supported by scientific arguments in order to be rational. After all, if scientism is rejected, it could be that there are many rational non-scientific reasons for beliefs. However, many take the critique of scientism yet further, to the conclusion that arguing about evidence of design and purpose in nature is not relevant to religious belief at all. Nor does the rejection of scientism imply that religious belief is automatically rational; after all, there are also philosophical arguments against the rationality of religious belief that can be appealed to by those who reject scientism.

Indeed, in contemporary discussion, the continued existence of natural theology (defined as the attempt to present evidence supporting religious belief) is sometimes met with incredulity. Have contemporary natural theologians not heard of Kant, Hume, Darwin and Barth? Nevertheless, natural theology has recently been undergoing a resurgence, as exemplified by volumes like the *Oxford Handbook of Natural Theology* (Brooke et al., eds. 2013) and the *Blackwell Companion to Natural Theology* (Craig and Moreland, ed. 2012).[4] Motivations for doing natural theology differ based on the type of natural theology to be adopted, but one commonality is the rejection of the most extreme forms of fideism, in which faith and reason are kept wholly separate (Vainio 2010).[5] The anti-evidentialists have valid points: religious faith is not reducible to science, and the rationality of faith does not require being able to present a convincing argument. Nevertheless, as McGrath (2004, 83) notes, "there is a growing feeling within the theological community that Barth's theology marks an over-correction" in its critique of natural theology. It seems that the linking of faith and reason allows for some goods that cannot otherwise be obtained, such as the pursuit of truth, the possibility of testing ideas about religion (and challenging misconceptions) in reasoned dialogue, and the value of the intellectual desire for God. It is desirable to avoid the danger of isolating the Christian faith in its own cultural ghetto, with no common ground or point of contact with the surrounding reality, and no way to criticize or support the rationality of belief by appeal to evidence (Kojonen 2020; Loke 2019).

According to the principle of epistemic fit, dating back to Aristotle (2000 [350 B.C.], 5), our methods of gaining knowledge about something should be adapted to the nature of that object.[6] For example, if we wish to gain knowledge about atoms, we should not expect to be able to do so without specialized instruments. Similarly, if we wish to gain knowledge about God, we should not necessarily expect that this knowledge should be gained using wholly the same methods as knowledge about atoms. Instead, as many in the philosophy of religion have noted, the possibility of this knowledge could be affected by factors that affect other personal knowledge.[7] Mikael Stenmark (1995, 325–327) thus argues that knowledge of God, like knowledge of a spouse's love, builds on lived experience, and the grounds of religious belief can therefore be much broader than can be communicated in a natural theological argument. Nevertheless, it is possible to treat both belief in a spouse's love and belief in God's love as hypotheses for the purposes of a conversation, even while being aware that these beliefs are not merely hypotheses. This is possible as long as the belief in question has at least some explanatory power.

Giving some space for natural theology also seems to be implied by the doctrine that God is the Creator of the entire cosmos, and that it has the God-given capacity to reflect his glory. While we may not want to say that we can "prove" the existence of God, or that the revelation of the "book of nature" (Juurikkala 2019) can be interpreted without the assistance of grace, it still seems desirable to say that the Christian faith can make sense of reality in a way that provides some intellectual support for it. Moreover, natural theology can be done in many ways, at least some of which share the concern for the centrality of Christ, and do not reduce faith to a matter of unaided human rationality. Proponents of natural theology also commonly share the idea that in contrast to critics like Barth, the Bible and the Christian tradition support the practice of some form of natural theology, and that the common Enlightenment-era critiques can be responded to (Kojonen 2017a).[8]

John Hedley Brooke (2002, 163–164) defines natural theology as "a type of theological discourse in which the existence and attributes of the deity are discussed in terms of what can be known through natural reason, in contradistinction (though not necessarily in opposition) to knowledge derived from special revelation." This is the usual understanding of natural theology, and there is support for this type of discussion from the Christian tradition (Swinburne 2004b, 536; Levering 2016b). However, demanding "knowledge" seems too stringent. For example, Swinburne, one of the most prominent defenders of natural theology in contemporary philosophy of religion, in his book *The Existence of God* comes to the modest conclusion that the overall evidence raises the epistemic probability of the existence of God to at least 50%. According to Swinburne (2004a, 2), while "reason can reach a fairly well-justified conclusion about the existence of God, it can reach only a probable conclusion, not an indubitable one." Swinburne himself believes that God's existence is more certain than this (particularly because of the evidence of religious experience and the evidence for Jesus' resurrection) (Swinburne 2003). Nevertheless, to also fit probabilistic arguments under the umbrella of natural theology, I prefer William Alston's (1991, 289) definition of natural theology as "the enterprise of providing support for religious beliefs by starting from premises that neither are nor presuppose any religious beliefs." This definition concentrates on evidential support for religious belief. It does not presuppose that these religious beliefs have been acquired through natural theology in the first place, that religious beliefs should be based solely on natural theology, or that natural theology should result in a proof. Using this definition, then, in the context of this volume, saving or salvaging the biological design argument would mean showing that biology's apparently purposeful order can provide at least some supporting evidence for religious belief, even for a person who accepts the veracity of mainstream evolutionary biology.[9]

As such, natural theology is compatible with rejecting the idea of a "neutral ground" from which reason can begin its inquiry into God. Utilizing the work of Alasdair MacIntyre 2016 among many other sources, Alister McGrath argues that there is no neutral "rationality" or a neutral "nature" from the point of which the truth of Christianity can be evaluated. Rather, for McGrath (2001, 2016, 2018), human perception of nature is always already based in some tradition. Nature itself is somewhat ambiguous about God, and human sin also hinders our reception of the natural revelation that does exist, he claims. Thus, the light of Christ must come first, and the full reception of natural revelation only afterwards.

Some might consider McGrath's approach to be a "theology of nature" rather than a natural theology. In Barbour's (1966, 415) terms, a theology of nature takes "the findings of science into account when it considers the relation of God and [humanity] to nature, even though it derives its fundamental ideas elsewhere." The fundamental insight here is that the question of discovering signs of the Creator in nature, so central to natural theology, is not the only important question in the relationship of religious "faith and wisdom" with science (McLeish 2014). McGrath also attempts this wider task of theological engagement with science. Yet the traditional task of natural theology is also part of his approach, since he argues that Christianity provides the "best explanation" for various features of reality. Thus, though some critics (e.g., Holder 2012) disagree, McGrath himself believes that his approach fulfills the apologetic function of traditional natural theology, rather than being of use only for those who already find themselves within the Christian faith. He claims that the explanatory power of Christianity is communicable, and thus "natural theology offers a comprehensive means by which theology may address the world, and engage in productive dialogue concerning the legitimation and consequences of belief systems" (McGrath 2001, 303). While McGrath's natural theology does not begin as an attempt to prove or justify Christian belief, but as a "faith seeking understanding," he nevertheless argues that the positive outcome of this seeking is "itself evidencing faith" (McGrath 2016, 176).[10]

McGrath's re-imagining of natural theology echoes Cardinal John Henry Newman's famous quip in critique of Paleyan natural theology: "I believe in design because I believe in God; not in God because I see design" (Newman 1974 [1870], 97; McGrath 2009, 128). However, for McGrath, faith does allow religious people to perceive nature as it really is, not simply as how they wish it to be. But it seems to follow from this that the effect of grace in allowing us to see design in nature cannot allow us to see something that is not really there. As John Lamont (2004, 12) points out, "No enhancement of cognitive capacities can enable them to go beyond what the evidence available to them can warrant them in believing; if a change to our cognitive capacities were to do this, it would not be an enhancement but a corruption of them."[11] Thus, if grace enables us to see evidence of design, this evidence should be really there, and should really provide support for belief in God. But this means that it should be consistent with McGrath's approach to assert simultaneously both that "I believe in design because I believe in God" and that "I believe in God (partly) because I believe in design."

There does seem to be something in the design argument that speaks to widely different cultures. The argument was used already in pre-Christian cultures, in Greek and Latin thought (Sedley 2007). A remarkable convergence with philosophical theism in the Hindu tradition can also be seen in that Hindu philosophers independently developed defenses and critiques of the design argument that are in many cases very similar to Western ideas (Brown 2008; Ganeri 2015; De Cruz 2014). Indeed, many contemporary thinkers argue that the possibility of natural theology and design arguments is rooted in the nature of human cognition. As Del Ratzsch (2010) puts it, "design thinking may be natural to our sorts of intellects." Thus, it seems credible that our common humanity also provides at least some amount of common ground, some "transversal rationality" (van Huyssteen 1999; Ratzsch 2009) that can be appealed to when discussing varying perspectives on design. Nevertheless, empirical studies do also make it clear that the perception of even human design is dependent, to some extent, on background information and prior beliefs in the existence of a designer. For example, people are far more likely to interpret abstract drawing as representative if they have information on the intentions of the artist (De Cruz and De Smedt 2015, chapter 4).[12]

In my own analysis of the debate on design arguments, I have also come to the conclusion that widely varying understandings of rationality are in play. When discussing design, it cannot even be assumed that everyone will agree on the reasonability of any intentional, personal explanations. Eliminative materialists argue that, even in the context of human action, such explanations are just unreliable "folk psychology" that should eventually be replaced with scientific talk (Ramsey 2019). For many observers (e.g., Cunningham 2010, chapters 5 & 6), eliminativism is a *reductio ad absurdum* of scientism—the refutation of scientism by showing that it leads to an absurd conclusion, the denial of the real effects of human consciousness and human persons on the world. Concepts of nature and the tendency to focus on the beauty or horrors of nature, and to see nature as fundamentally good or fundamentally flawed, are also culturally influenced (McGrath 2001; Garvey 2019).

To summarize, in my understanding, natural theology can well begin as a faith-seeking understanding, but in showing the resonance and coherence of the evidence with religious belief, it should also attempt to provide a defense of the rationality of belief, insofar as this is possible across traditions. This, then, implies that salvaging the biological design arguments can proceed in two stages: (1) rehabilitating the idea of biological design as a sign of the Creator when interpreted from a theistic or Christian starting point, as part of a theology of nature, and (2) showing how biological design can function as evidence of the Creator more broadly, in a natural theology aiming for dialogue between theists and non-theists. Particularly in this latter approach, purpose and design should not be simply thought of as something theologically "read into" nature (with the equally rational possibility of an atheistic reading), but also as something that is "read off" nature's order (MacSwain 2013, 197). But there is still another objection against the value of design arguments: the apparent anthropomorphism of describing God as a designer.

God the Designer?

Some proponents of design arguments, such as those in the ID movement, would not describe their arguments as theological at all. In the early stages of the movement, when "intelligent design" was not yet widely used as a term, even Phillip Johnson, the visionary leader of the movement, initially constructed ID's argument as a comparison of supernatural creation and Darwinian evolution as explanations (Johnson 1993, 1995).

He argued that evolution was based on a methodological naturalism that assumes that no supernatural creation has taken place a priori, and thus rules out the true theistic explanation before consideration of the evidence.[13] An early collection of articles by ID supporters was even titled "The Creation Hypothesis" (Moreland ed. 1994), and another "Mere Creation" (Dembski ed. 1999). However, later ID thinkers have emphasized the religious minimalism and scientific nature of their design argument (Kojonen 2016a, 89–92). Johnson himself later joined in this chorus, writing that "My personal view is that I identify the designer of life with the God of the Bible, although intelligent design theory as such does not entail that" (Johnson 2007).

ID's minimalistic design argument raises the ire of Hanby (2016, 166n51), who argues that "if your conception of the God–world relation is such that God can be replaced by a space alien without loss, there is probably something wrong with both." This may be unfair to proponents of ID, since their design argument does not claim to provide a full account of the God–world relation. However, in contrast with ID, Paley (2008 [1802], XXVI) emphasized that the Creator of life possesses all of the traditional omni-attributes of the Christian God. But Hanby also finds Paley's concept of God overly anthropomorphic. According to Hanby, the faults of Paley's theology are evident already in the analogy of the watch:

> Paley ambles across his imaginary heath in the opening pages of the *Natural Theology* undisturbed by his surroundings. The heather, the stones, and the blue sky above elicit no commentary from him. The vista ostensibly opening up before him goes without mention. Paley finds nothing remarkable in the utter contingency and irreducible uniqueness of all that exists or even in the fact, which Balthasar called 'astonishing beyond measure,' that it is not just someone but *he* who is able at that instant to carry on such an exercise in his imagination (Balthasar 1991, 615). It is only when he happens upon his hypothetical watch with its mechanical "contrivances" adapted to a common function that he is moved to curiosity. (Hanby 2016, 153)

In order to properly approach the Christian doctrine of creation, Paley should have felt wonder already at the basic mystery that there is anything at all, as well as the mystery of his own conscious existence. Hanby thus claims that if Paley had understood Christian doctrine, he should also have realized that the opposition between natural causes and creation is unnecessary. Hanby (2016, 163) admits that "one can find appeals to the artisan analogy for creation and even justifications for divine goodness from observed utility scattered throughout the tradition." However, for Hanby, it is always surrounded by the proper theological context, in which the analogical nature of these descriptions is clear, whereas Paley's argument uses the terms in an anthropomorphic way. Thus, Paley's God is not truly God, but simply a powerful and smart engineer.

The critique of Paley's and ID's language of design as anthropomorphic is directly relevant for my argument. If describing God as a designer is misguided up to the point of idolatry, what point is there then in trying to salvage the biological design argument? It does seem clear that Paley's design argument indeed presents only a partial picture of God as Creator, and insofar as others have taken Paley's thought as the zenith of Christian theology, this is unfortunate. Hanby may also well be correct in arguing that the anti-theology of many evolutionary biologists has been influenced by such a narrow conception. Yet I think Hanby also goes too far in his critique of Paley and ID. Let us assume that Hanby is correct in his judgment that Paley was not in fact moved to consider God based on the existence of consciousness and the world as such, but only when encountering the ordered complexity of organic beings.[14] Nevertheless, this would not mean that Paley's particular disposition is a necessary feature of the design argument itself. We can well imagine a Paley who would have been astonished by the sheer mystery of being, but also astonished by the contrivance of living organisms.

Hanby is correct that Christians understand God as much more than a designer, and in classical theism as the "fullness of being." Yet in the classical understanding, God is also not less than a designer. All the properties that we value in the case of humans, such as the ability to act purposefully, partially reflect the fullness of God. In this sense, it does not seem that calling God a designer is any more problematic than the use of other human expressions, such as describing God as "loving." Biblically, our human love is not the same as divine love, "for my thoughts are not your thoughts, neither are your ways my ways, declares the Lord" (Isaiah 55:8), yet there is a clear connection between the two: "anyone who does not love does not know God, because God is love" (1. John 4:7). In traditional theological discussion of divine attributes, it has been emphasized that we should move between "cataphatic" and "apophatic" language, affirmations and denials (Vainio 2020). God is love (the affirmation), but our human experience of love is incomplete, just a dim reflection of divine love (the denial).

There is debate here to be had on what exactly is meant by analogy, and whether there must always be some shared "univocal" meaning when a term like "love" is applied to both God and humanity (McCord Adams 2014; Turner 2004; Williams 2005). The arguments for at least some univocity are difficult to respond to, for otherwise theological language risks losing all meaning. However, Thomists who emphasize the analogical nature of theological language must still do justice to the scriptural descriptions of God, which describe God acting purposefully and toward goals. If such biblical descriptions can be harmonized with Aquinas' type of classical theism, as Eleonore Stump (2016) has argued, then I do not see why the same could not apply to the talk of God's intentions and purposes in design arguments. William Abraham (2019) similarly argues that describing God as an agent does not require assuming that God is akin to human agents in all respects. The word "design" may indeed bring to mind God as merely an engineer, and so it needs to be balanced against the whole of the doctrine of creation, in order to remind us of the dependency of all being on the Creator. However, the doctrine of creation does also affirm that God is the orderer of the cosmos, and this is even an aspect which is emphasized in the creation narrative in Genesis 1—to the extent that some have even argued against seeing the doctrine of creation out of nothing in the text (however, see Levering 2017; Craig and Copan 2004; Burrell ed. 2010).

For Paley, the design argument did indeed occupy a central role in justifying his understanding of nature as God's temple. Recall what Paley argued regarding the usefulness of the biological design argument: "The change is no less than this, that whereas formerly God was seldom in our thoughts, we can now scarcely look upon anything without perceiving its relation to him" (Paley 2008 [1802], XXVII). As noted, I agree that there is definitely much more in the cosmos and in our own existence that could have also elicited thoughts of God. Hence, I do not think that reaffirming the reality of biological design by itself needs to create such a large change in our thought, so that on its own it would need to fully transform our view of the cosmos from a secular cosmos to a temple. This is unnecessary if we have properly understood the broad basis of the doctrine of creation. However, this does not mean that the meditation on biological design and divine glory in it could not also contribute to this kind of spiritual interpretation of nature, as one among many other considerations.[15]

I have now argued in favor of the thesis that evidence of design can be relevant for religious faith, and that science need not be our exclusive guide to reality. The different understandings of natural theology, and the discussion of limit questions, will be relevant in later chapters, when I go on to develop my account of the compatibility of the biological design argument and evolution, and how it might fit within a natural theology that also takes into account the effect of our broader worldview. I will now move on to another foundational issue, namely to consider how my approach relates to other ways of understanding the relationship between evolution and creation.

2.2 On the Compatibility of Evolution and Creation

The Assumption of Conflict

On the popular level, it is common to understand evolution and creation as incompatible, and this idea is even repeated by some academics. For example, evolutionary biologist Douglas Futuyma (1983, 13) claims that "some shrink from the conclusion that the human species was not designed, has no purpose, and is the product of mere mechanical mechanisms—but this seems to be the message of evolution." Such theological interpretations of evolutionary biology have been proposed by many other evolutionary biologists as well (Haught 2004). Following this interpretation of the theological significance of evolution, many proponents of Intelligent Design also hold that while evolution and creation may be barely logically compatible, there is nevertheless significant tension between these ideas. For example, my countryman, biochemist Matti Leisola (2017, 163), states that "Darwin and his followers explicitly claim that chance variation and the law of natural selection have created all species of living organisms. If so, design is ruled out. In contrast, if design is the cause of the creation of species, then chance and law are ruled out as the creators of species."[16]

In contrast, this study begins from the assumption that the doctrine of creation is fundamentally compatible with evolutionary biology, even when natural selection working on chance mutations is understood to be the central evolutionary mechanism. This assumes that it is possible to separate the "dross" of evolutionist metaphysics from the "gold" of evolution as a scientific theory, as Ted Peters and Martinez Hewlett (2003, 22) and many others (e.g., Plantinga 2011) argue. In my view, this is not a "compromise" of the doctrine of creation to fit evolutionary ideas, as creationists would have it. Rather, the compatibility of creation and evolution arises from the foundational theological core of the doctrine of creation. All natural processes are dependent on the divine act of creation, and so cannot be in contradiction with it. Thus, the in principle compatibility of creation and evolution would hold as a theoretical truth even if evolutionary theory itself were overturned.

In Genesis 1, God creates the universe and commands the earth to bring forth the various species of plants and animals. Commenting on this in *The Literal Interpretation of Genesis*, Augustine formulated the idea that God created the species through rational principles, rather than creating them directly (Levering 2016a; McMullin 2009). Aquinas defended the idea that God is the primary cause of all creation, working through secondary causes within his creation (Kretzmann 1999). Reformer Martin Luther, explaining the meaning of the first commandment, similarly sums up an important underlying principle: created beings can function as the hand, channel and means through which God continually gives to us (Luther 2016 [1526], WA 30.I, 132–139; Swanke 2014). This is not to deny that these influential theologians also believed that God acts miraculously within his creation. But it does show that the general idea of creating through a natural process does not contradict the meaning of the doctrine of creation.

As stated in the Catholic Church's document "Communion and Stewardship: Human Persons Created in the Image of God" (Vatican International Theological Commission 2004, 69), "neo-Darwinians who adduce random genetic variation and natural selection as evidence that the process of evolution is absolutely unguided are straying beyond what can be demonstrated by science. Divine causality can be active in a process that is both contingent and guided. Any evolutionary mechanism that is contingent can only be contingent because God made it so. An unguided evolutionary process—one that falls outside the bounds of divine providence—simply cannot exist." The document thus argues that any idea of a necessary opposition between evolution and creation arises either from a misunderstanding of the doctrine of creation, or from an overreach in the interpretation of scientific results.

Divine Action and Chance

A central issue here concerns the definitions of randomness and contingency in evolution in relation to divine action. I will have occasion to discuss this often in the coming chapters, but some remarks are already due here. George Ellis (2018) sums up three different types of chance: (1) chance as epistemic ignorance of the deterministic processes that are actually responsible for the outcome (as in roulette or in lotteries), (2) chance as the juxtaposition of deterministic but unrelated causal trajectories (as in car crashes, or when two people meet by accident), and (3) real ontological indeterminism, as found in the Copenhagen interpretation of quantum mechanics.[17] Further distinctions could be made, particularly with regard to different understandings of the relative importance of contingency in nature (Giberson, ed. 2016; Fergusson 2018, chapter 4). As Ulanowitz (2016, 363) notes, "One thus sees that Monod's crisp dichotomy between 'chance and necessity' is a gross oversimplification. Instead, there exists an entire spectrum of contingencies ranging from radical chance at one extreme to blind chance, conditional probabilities, propensities, and finally to deterministic phenomena." The aforementioned "Communion and Stewardship" document would allow for all of these types of chance to be compatible with divine providence. However, what biologists primarily have in mind is the second point: the juxtaposition of unrelated causes.

According to the standard view, the mutations that provide the raw material for evolution are random in the sense that they are generally unrelated to the needs of the organism, with harmful, neutral, and beneficial mutations as possibilities that are then weeded by natural selection. As philosopher of biology Elliott Sober (2011, 192) writes, there "is no physical mechanism (either inside organisms or outside of them) that detects which mutations would be beneficial and causes those mutations to occur."[18] The reference to a "physical mechanism" leaves out discussion of God, but Sober is clear that in evolutionary theory, the scientific explanation cannot depend on God. However, he also argues that evolution, as a probabilistic theory, cannot rule out divine guidance: "The theory is consistent with there being *hidden variables*, natural *or* supernatural. […] Theistic evolutionists can of course be deists, holding that God starts the universe in motion and then forever after declines to intervene. But there is no contradiction in their embracing a more active God whose post-Creation interventions fly under the radar of evolutionary biology. Divine intervention isn't part of science, but the theory of evolution does not entail that none occurs" (Sober 2011, 196–198). At most, evolutionary science can say that there is a highly plausible cause for the origin of species that does not require supernatural interventions in order to work.[19]

Barbour (1997, 239–240) outlines three different broad views of theistic evolution, and his categories provide a good starting point for understanding different defenses of the compatibility of evolution and creation. In the first view (1), God controls the evolutionary events that appear to be random from the point of view of science. Though this supervision is undetectable through scientific means, it is still be believed to be there for theological and philosophical reasons. For example, Robert J. Russell (2019) defends a theory of non-interventionistic divine action in which God directs the process of evolution through influencing events on the quantum level, resulting in divine guidance of evolutionary mutations, although these appear to be random from the point of view of science.

In the second view (2), it is thought that God directs evolution through setting the process in motion in a very particular way, constructing the environment and the preconditions of the process so that his goals are guaranteed, even with randomness included in the process. As Tracy (2013, 464) puts it, "God acts always and everywhere as the transcendent creative ground of the world's existence, and God's purposes are realized in and through the operation of created causes in the natural order."

In the third view (3), God gives the evolutionary process a great deal of freedom, influencing events but not absolutely controlling the direction of evolution, instead opting to mystically draw the creation toward his love. This view is traditionally related particularly to process theism, which abandons aspects of traditional theism, such as omnipotence. However, many have also argued for the compatibility of the autonomous evolution of creatures with traditional theism, particularly as a response to the problem of evil (Southgate 2008; Sollereder 2018). In my argument in this book, the constraints of the evolutionary process will play a significant role, but the argument also aims to leave room for the freedom of creatures and for divine action beyond setting things in motion.

Combining belief in creation and evolution also requires discussion of more specific matters of doctrine and biblical interpretation, such as the matter of the historicity of Adam and Eve, as well as the relationship between sin and death. For considerations of space, I will simply have to lay these important matters aside here and refer the reader to other works (Rosenberg et al. ed. 2018; Torrance and McCall ed. 2018a, b; Cavanaugh and Smith ed. 2017). However, I invite the reader to proceed with me, for the sake of the argument, in assuming that these problems can be solved, or to just leave them aside for now. This will then allow us to focus on the specific question of the compatibility of design arguments and evolution.

Beyond Mere Compatibility

The mere compatibility of evolution and creation still leaves important questions unanswered. Centrally for this book's argument, the question of the *overall compatibility* of creation and evolution is distinct from the question of the *revelatory potential* of biological creation. Thus acceptance of compatibility does not yet mean that evolution has no epistemic effects on some religious beliefs. Recall Ruse and Dembski's argument that evolution gives us a new, godless understanding of the world, and explains away the evidence of biology, thus removing a central reason for belief in the Creator. Ruse (2018, 282) himself also seems to think that "after Darwin, the creation no longer sings" of its Creator, though he believes that religious belief in creation is compatible with evolution. Dembski (1999, 111–112) argues that "For the hard-core naturalist, theistic evolution at best includes God as an unnecessary rider in an otherwise purely naturalistic account of life. Thus by Occam's razor, since God is an unnecessary rider in our understanding of the natural world, theistic evolution ought to dispense with all talk of God outright and get rid of the useless adjective *theistic*. This, at any rate, is the received view within the Darwinian establishment." As I have noted elsewhere (Kojonen 2016a, chapter 10), many proponents of ID similarly see scientific support for belief in the purposiveness of biology as essential for the credibility of religious belief in creation. Dawkins' (1986, 6) famous argument that Darwinism makes an "intellectually fulfilled atheism" possible is also about the purported epistemic effects of evolution.[20]

However, though I agree with the general idea that the revelatory potential of biological creation also needs to be addressed, it still seems to me that an overly narrow understanding of religious rationality often underlies such claims—something that Ruse also acknowledges elsewhere (Ruse 2001). One could get the impression that the options of the religious believer are either an extreme fideism, in which religious faith is based on a "leap of faith" without any basis in evidence, or a hardline evidentialism, in which religious faith is only rational insofar as it is grounded in publicly available, even scientific, evidence. This would be a kind of religious adoption of the ideology of scientism. However, in fact, the options available are much broader than that—and very few religious thinkers would fit into either extreme (Vainio 2010). Insofar as evidence is relevant for religious belief, the apparent teleology of biology is hardly the only evidence to be taken into account, or even the most central evidence. Beyond questions like "How is it possible for the parts of the human eye and brain to have been so exquisitely ordered toward the purpose of granting the ability to see?", we can further ask "How is it possible for anything contingent to exist at all?", "How is it possible for conscious minds to understand the cosmos?", and "How can there be good and evil?", questions which invite natural theological arguments. Ruse's conclusion that after Darwin "nature no longer sings of its Creator" would also need to be justified against the broader discussion in the traditions of philosophy of religion and natural theology (e.g., Craig and Moreland, eds. 2012; Pruss and Rasmussen 2018; Loke 2017; Evans 2014) not merely against the biological design argument. Moreover, various testimonial, experiential, and other reasons could also provide us with non-inferential reasons for belief (Abraham and Aquino ed. 2017). All such reasons can also be criticized, but this does not mean that they are without value.

Blaise Pascal famously argued that "in faith there is enough light for those who want to believe and enough shadows to blind those who do not." Defending this principle, C. Stephen Evans (2012) argues that theistic cosmological, teleological, and moral arguments build on "natural signs" that are present in all of creation, and that provide intuitive grounds for belief in God ("enough light"). However, Evans also argues that all of the signs are resistible (so that there are "enough shadows"), and that this should be so in order to make human freedom and moral testing in the present creation possible. In my view, though the theistic arguments building on such signs can indeed be criticized, natural theologians have nevertheless succeeded in providing us with many defensible arguments for the existence of God. Thus, even accepting the importance of publicly available evidence for supporting theistic belief, theistic evolutionists do not need the biological design argument in order for theistic evolutionism to be intellectually defensible.

2.3 Perceiving Design in Nature

Design Arguments and the "Theist on the Street"

Given the compatibility of evolution and creation, what then is the importance of saving the biological design argument? Here is how I see the issue. Following Evans' (2012) argument about natural signs, it is beneficial for an argument of natural theology to build on broadly accessible phenomena. Biological design is observable in our own bodies and in the order of life all around us, and thus satisfied this condition better than many design arguments that are more dependent on knowledge of theoretical physics, for example.[21] It seems that something about the nature of biological organisms leads the human mind easily to the idea of design. As I will soon describe, many have defended the idea that design beliefs are not necessarily primarily the result of arguments, but rather the result of intuitive design detection comparable to perception. This "naturalness" of design beliefs has been used both to support the rationality of these beliefs and to undermine them, but in my view it in itself does not yet prove the matter either way. However, the common appeal of the idea of biological design does show that it fulfills the criteria of a good natural sign of the Creator, as described by Evans (2012, 2018). If the belief that biology manifests divine purpose can be defended as rational against possible counterarguments, such as evolutionary explanations, then this means that we can recognize the ordinary "theist on the street" (Shaw 2019) as rational in perceiving biological nature to be created. A successful salvaging of the biological design argument, or even just the refuting of evolution as an objection against design beliefs, would allow for attaining the valuable goal of defending the rationality of this kind of intuitive religious interpretation of biological order.

In contrast, if our intuitive design detection is shown to be wrong about biology, this might also weaken our trust in our capacity to detect design elsewhere. Dawkins (2006, 139) thus argues that "a deep understanding of Darwinism teaches us to be wary of the easy assumption that design is the only alternative to chance, and teaches us to seek out graded ramps of slowly increasing complexity. […] After Darwin, we should feel, deep in our bones, suspicious of the very idea of design." Dawkins' argument is not convincing. Being wrong in detecting design in one area does not mean that we could not detect it elsewhere. The fine-tuning argument could be convincing even if the biological argument is not. Nevertheless, in a limited sense, Dawkins is right: if the perception of design in biology is fundamentally in error, this would still be a blow against approaches that emphasize the *prima facie* reliability of such commonsense reasoning in natural theology.

The salvaging of the biological design argument would also provide a theologically more robust and consistent version of theistic evolutionism. Wahlberg, commenting on Nancey Murphy's arguments on perceiving design, makes the case that Christian theists have good reason to at least attempt to defend the possibility that the human perception of purpose in biological nature is truth-tracking. Murphy, in e-mail correspondence with Wahlberg (2012, 188), had argued that:

> Before Darwin, Christians thought they were directly seeing, without intervening theory, God's intentions in nature. Now we know that what appears to be design is not the direct manifestation of design (as is pain behavior a direct manifestation of pain) but instead have to recognize that it only appears to be a direct manifestation of design. Only indirectly, via THEORETICAL (theological) knowledge, can we still say that it is, in a different sense, the product of intention.

As Wahlberg notes, the situation described by Murphy is strange. Supposing that God has created living organisms purposefully, then the Christians who thought they were seeing divine purpose were correct purely by luck but without valid reasons. This would mean that reality resembles the famous Gettier paradoxes (Gettier 1963; Pollock 1986, appendix). It is, Wahlberg says, as if you think that there is a yucca tree in the room. However, what you see is just an illusion created through mirrors. Nevertheless, behind the illusion there really is a yucca tree in the place where the illusion leads you to expect it. However, you did not see that tree, so you came to the right conclusion purely by luck, and thus did not possess justified true belief.

Similarly, Murphy seems to affirm both (1) that God intentionally brought about that complex organisms exist and (2) that these organisms appear to have been intentionally brought about. Nevertheless, for Murphy the appearance of design does not give us reason to believe that organisms were actually purposefully created. Rather, we can know this only through inference from our other religious beliefs. But it would be desirable for our understanding of reality to avoid such paradoxes, and to avoid explaining correct beliefs as the result of luck. Theists seem to have an alternative explanation: perhaps God wanted people to have such "seemings," and set up the evolution of cognition to make this actually happen. Moreover, biology's order could be actually revelatory of the Creator, even in an evolutionary cosmos. It seems that Christian theists should at least be interested in exploring these possibilities in order to avoid such a paradoxical and counterintuitive situation.

Design Discourse Rather Than Design Arguments?

Due to the clear importance of experience in grounding design arguments, many have recently argued that design beliefs should be thought of as akin to perceptual beliefs, rather than primarily as inferential beliefs. Alvin Plantinga (2011) thus claims that it might be profitable to speak of "design discourse" rather than design arguments. Ratzsch (2003), Mullen (2004), and Evans (2018) have also proposed similar approaches. Wahlberg (2012, 195) summarizes the resulting understanding: "I know that nature is creation because (so I contend) I *see* that it is (i.e., I see that biological structures are expressive of intent and intelligence)."

This basic approach can be traced back to at least William Whewell (1794–1866) and Thomas Reid (1710–1796) (Ratzsch 2003, 125). Whewell (1834) argued that "when we collect design and purpose from the arrangements of the universe, we do not arrive at our conclusion by a train of deductive reasoning, but by the conviction which such combinations as we perceive, immediately and directly impress upon the mind." Reid's (2002 [1785], V) basic point was that the same is true of the vast majority of our beliefs: typically they are produced by capacities, rather than arguments. In contrast to those who claimed that we should be able to defend our beliefs by reference to evidence before considering these beliefs to be justified, Reid adopted a kind of "innocent until proven guilty" assumption for epistemology (Chignell 2018). According to Reid, we simply have to begin from the standpoint of trust in our cognitive faculties. If we do not do so, then we will ultimately be unable to ground any of our beliefs, for there is no way to justify the reliability of our faculties that does not already rely on them (Alston 1991).

Reid also applied this to design detection. According to Reid, the capacity to detect intelligent causes is a basic human capacity that we need in order to be able to operate at all in human society, and even to recognize intelligence in our immediate family members. The same capacity that allows us to perceive that other humans have minds also allows us to perceive that nature is created, with a comparable level of certainty. Reid does allow for design arguments to further reinforce the reliability of this initial perception, but claims that such arguments are not necessary for belief.

Currently, the primary defenders of Reidian epistemology come from the traditions of *reformed epistemology* and *reliabilism*. Reliabilists, such as Plantinga, argue that the reliability or "warrant" of our beliefs does not primarily arise from internal justification in the form of reasons that we can present for our beliefs, but from reliable faculties. For example, if I see a tree in front of me, I am justified in believing that there is a tree there. Plantinga further argues that beliefs about God can also be the kind of "basic beliefs" that can be believed without argument, as basic beliefs that are formed by our faculties in religious experiences, for example. As long as these basic beliefs are not defeated by a counterargument or other reason, then the theist can reasonably hold to them. For instance, if I remember that I am wearing virtual reality goggles, then I have a defeater for my belief that I am perceiving a real tree in front of me. Plantinga argues that properly basic beliefs might also receive further support from evidence and arguments, and so his defense of design discourse is also compatible with developing design arguments (Mascord 2007).

There are many critiques of reliabilism. For instance, it has been argued that it is overly permissive, granting the status of rationality even to beliefs that we would want to reject as irrational (Moon 2016). Other problems are related to, for example, the difficulty of defending complex religious beliefs as properly basic (Baker-Hytch 2018). A possible counterargument focuses on the nature of perception. One critique has been to deny that theism could be a basic belief. Similarly, a critic of design discourse could claim design beliefs are more like interpretations of nature than perceptual beliefs. As Paley (2008 [1802], I) himself admitted, the ability to perceive the purpose of the mechanisms of a watch requires some prior knowledge and a detailed study of the parts. And while Darwin did state that he sometimes experiences the "overwhelming force" of the idea of design, he continued by saying that at other times, this sensation seems to go away (Argyll 1885). This is in contrast to the perception of trees, which are not as dependent on the observer's mental state. However, as McGrath (2008, chapter 5) notes, psychological theories on perception have long emphasized that human perception is always influenced by the schemas we interpret our sensations by, and even by our affective orientations. Thus, insofar as all perception is dependent on concepts, the dependence of design perceptions on concepts presents no reason by itself to be suspicious of them.[22]

Happily, it seems to me that intuitive design beliefs do not necessarily need to be classified as perceptual beliefs in order to be experientially grounded basic beliefs. In contrast to reliabilists like Plantinga (who defend an "externalist" epistemology), internalists claim that our beliefs should be based on reasons that are internally accessible to us. But this also does not require treating design beliefs as inferential. Internalists can also hold that simply "seemings," such as "it seems to me that there is a tree in front of me," should already be taken as a good reason, or as good evidence, for a person to believe that there is indeed a tree in front of them (Dougherty 2018). Swinburne (2011, 2012), for example, has defended a "principle of credulity," according to which if it seems to a subject that x is present, then probably x is present, unless there is counterevidence. As Swinburne (2012, 43) states, a compelling justification for this is that "if we cannot start our epistemic journey from our basic beliefs, we cannot start it at all, and total scepticism would follow." In an understanding of evidentialism in this vein, if nature, or something within nature, seems designed to a person, then he or she can take this seeming as evidence of the reality of design. Whether we regard this as a compelling reason to believe in design will then depend on the strength of the seeming, and whether design fits in with our understanding of the total evidence. While it is true that experience often also disproves our intuitive ideas, including our "seemings" (Stitch 1985; Burns 2017; Peels et al. ed. 2020), without assuming their basic reliability we cannot even get started in our inquiry into the trustworthiness of our faculties in particular cases.

Hence, we have here three main ways to express how design beliefs are generated. First, such beliefs could be understood as the products of conscious arguments, based on certain evidence in nature. Second, these beliefs could be the products of strong "seemings," which we intuitively infer as evidence of design. Third, these beliefs could be akin to perceptual beliefs. There are yet more possibilities: for example, we could interpret the significance of design as a sign of design by using semiotic theory. In this way, we would see apparent design as a sign of design akin to the way we see smoke as a sign of fire.[23] In practice, design beliefs can actually be supported by both inferential and non-inferential reasoning: a design argument in combination with the initial perception of design (von Wachter 2014, 60).

Ratzsch (2003) has argued extensively for the thesis that human design beliefs are typically not primarily the result of arguments, but are in some way non-inferential. I will summarize three core reasons here. First, in the course of our regular living we detect intentional activity every day, and we typically do this intuitively and almost instantly, without the need for extensive analysis. It is true that sometimes detecting design requires more work—for example, when trying to find the function of an obscure artifact, or collecting the evidence of a carefully concealed crime—but these situations are not typical. Here it can be objected that our intuitive design detection might still be based on tacit, unstated, and unconsidered inferential principles. However, as Ratzsch (2003, 135–136) points out, design arguments do in practice gain at least some of their appeal from their resonance with these intuitions, and many defenders of the argument have accordingly felt it unnecessary to analyze its core notions, such as "design" and "purpose." But it would still seem difficult to argue that no further analysis of design detection is beneficial beyond "it is perception." At the very least, we could analyze what kind of situations cause us to strongly perceive design, and what kind of situations do not (Mullen 2004). This would then allow analysis of the relevant properties, and whether we can justify trusting in our perceptions of design in our ordinary lives, but not in the case of nature.

Second (2), we have experiential testimony from both proponents and critics of design to the effect that the idea of design feels almost forced on them by the evidence, in a manner closely resembling perception. As Wahlberg (2012, chapter 6) notes, in descriptions of the intuitive force of the idea of design, the observer of nature is often described in passive terms, the idea being that the conclusion that nature is designed is almost forced on the observer by his perceptual capacities. Immanuel Kant, though critical of all arguments as a basis for religious belief, nevertheless writes that

> No one can have such a good conceit of his insight as to wish to assert definitely that, for example, the most admirable conservation of the species in the plant and animal kingdoms, whereby each new generation re-presents, every spring, its original, anew and undiminished, with all the inner perfection of mechanism and (as in the plant kingdom) even with their delicate beauty of color, without the forces of inorganic nature, otherwise so destructive, in the bad weather of autumn and winter being able to harm their seed at all in this respect—no one, I say, will assert that this is a mere result of

> natural laws; no one, indeed, can claim to comprehend whether or not the direct influence of the Creator is required on each occasion. (Kant 1960 [1793], A116)

The idea is even expressed in David Hume's *Dialogues Concerning Natural Religion* (2001 [1779]), considered by many to be the classic critique of all design arguments. The character Cleanthes, defending the design argument, asks us to "Consider, anatomize the eye; Survey its structure and contrivance; and tell me, from your own feeling, if the idea of a contriver does not immediately flow in upon you with a force like that of sensation?" (Hume 2001, III). It might be objected that this character does not reliably represent Hume's own views, but is it credible to think that Hume would have put these words into Cleanthes' mouth, if he did not himself also feel the pull of this intuition? In any case, the sentiment is also affirmed later by the character Philo, who is usually considered to most closely represent Hume's own views. Philo claims that "a purpose, an intention, or design strikes everywhere the most careless, the most stupid thinker; and no man can be so hardened in absurd systems, as at all times to reject it" (Hume 2001, XII).

Darwin, while rejecting Paley's version of the design argument, wrestled with the same experience of the "overwhelming force" of the idea of design (Argyll 1885, 244; Darwin 1887, 316). Describing the state of contemporary evolutionary biology, historian of science Timothy Lenoir (1992, ix; Ratzsch 2003, 135) claims that "Teleological thinking has been steadfastly resisted by modern biology. And yet, in nearly every area of research biologists are hard pressed to find language that does not impute purposiveness to living forms." The trend has been documented by many authors (Riskin 2016), and examples are not difficult to find. For example, Crick (1989, 30) writes that "organisms appear as if they have been designed to perform in an astonishingly efficient way, and the human mind therefore finds it hard to accept that there need be no Designer to achieve this." For Crick, it is only the advancement of science which allows us to now recognize that the idea of design is wrong: "Had I been living 150 years ago I feel sure I would have been compelled to agree with this Argument from Design" (Crick 1989, 24). In the debate over ID, atheistic critics of ID commonly agree with the intuitive appeal of the idea of design, though they then go on to argue that the totality of the evidence should lead us to reject it. Dawkins (1986, 2006), for example, builds his critique of design arguments in this manner.

Third (3), recent studies in the cognitive sciences also support the idea. There is much evidence to show that humans are not born as "blank slates," but that our cognition is built in a way that allows us to recognize other intelligences and interact socially (Clark and Barrett 2011; Visala 2011; De Cruz and De Smedt 2015, chapter 4). Many postulate that these same processes are also crucial in the formation of religious beliefs. In particular, postulated automated cognitive "modules" responsible for the detection of other minds (the *Theory of Mind Module*, ToMM), the agency detection module, and our innate capacity for probabilistic reasoning are argued to be necessary for our beliefs regarding divine agency. Thus the intuitive power and commonality of design arguments may be (at least to some degree) rooted in the cognitive structure of the human mind. However, cognitive scientists have differences regarding the precise relation between our cognitive makeup and our religious beliefs. Some argue that our cognitive structure makes us "born believers" (Barrett 2012) or "intuitive theists" (Kelemen 2004). Others argue for the more modest thesis that teleological beliefs concerning nature are cognitively natural or easy to form, but still require cultural support, particularly when speaking of full blown theistic beliefs (Bloom 2009).

In my view, these reasons do lend support to the idea that design beliefs have a strong intuitive component, and that such commonsense beliefs can be reasonable when no sufficient counterargument exists. This conclusion holds whether these design beliefs are understood as akin to perceptual beliefs, or simply strong "seemings." This does not mean that seeing nature as designed is not also influenced by one's cultural situation, or that the intuitive appeal of the idea is the only reason why people believe in design. There is also reason to think that testing such commonsense beliefs through argumentative discussion is also valuable. But there are also arguments against the reliability of our cognitive capacities in the case of design arguments. I will now turn to this.

Debunking Arguments

Cognitive scientists of religion and philosophers do not agree about the implications of their results for religious rationality (Visala 2011; Launonen 2018) Some have argued that in providing a natural, scientific explanation for religious belief, the theories of the cognitive sciences debunk religious ideas of the origins of belief. In this understanding, divine revelation and natural explanations are presented as competing explanations for the same phenomena of religious belief (Wilkins and Griffiths, P.E. 2013). Others, however, argue that these results support the old religious idea of a God-sense (*sensus divinitatis*), whose evolution was intended by God, though the *sensus divinitatis* should not be directly identified with some feature such as our agency detection module. Thus, no explanatory competition is necessary (Clark and Barrett 2011; Jong 2013; Henriksen 2015). Some argue that religion and belief in God are such complex phenomena that the postulated cognitive mechanisms can at most explain only a part of religion, not most of it (Leech and Visala 2011). However, design beliefs are among the parts that might be most affected, since they are, in this understanding, directly linked to the cognitive modules responsible for theory of mind and agency detection. Here the arguments of the cognitive sciences of religion form the basis of a debunking argument, which would function as a reason not to trust our perceptions or "seemings" concerning design in nature.

Guy Kahane (2011) lists the basic premises of evolutionary debunking arguments as follows:

1. S's belief that p is explained by X.
2. X is a process that does not track the truth.
3. Therefore, S's belief that p is not justified.

In the context of design beliefs, if the cognitive processes producing design beliefs, particularly supernaturalistic design beliefs, can be shown to be generally unreliable, then this would undercut design discourse. As De Cruz and De Smedt (2015, 195) summarize, assuming that such a debunking argument is successful, "the intuitions that lead us to infer a designer of natural objects have been shaped by, among others, our evolutionary history as designers and users of artifacts. While the evolved design stance does not cast direct doubt on theism, it can provide an undercutting defeater for the original evidence, that is, our perception of design and our inference to a designer."

However, it seems to me that none of the existing arguments for premise 2 (that our design detection processes are not truth-tracking) is successful. The most powerful such objection would be if evolutionary explanations undercut design beliefs, but that is precisely the point at issue in this book. In addition to evolution, the most often quoted argument for the unreliability of our intuitive design beliefs is the "hyperactiveness" of our agency detection originally proposed by Barrett (2004, 31). It seems plausible that the lives of our ancestors may have often depended on the ability to recognize even slight marks of the nearby existence of a predator, such as a tiger. But the claim is that survival would be affected more by false negatives than by false positives, and that therefore it would be evolutionarily beneficial for our agency detection to become hyperactive, even if this would lead us to false beliefs. Thus, the sound of leaves brushing together or the branch of a tree moving would cause us to feel alarmed or to look for an agent responsible for the sound, even if it were caused only by the wind. Arguably, falsely believing in supernatural agents would also not harm survival. Wilkins and Griffiths (2013, 143–144) conclude from this that our agency detection is not reliable, and they claim that "people believe in supernatural agents which do not exist for the same reason that birds sometimes mistake harmless birds passing overhead for raptors." This has led to much discussion about the alleged "promiscuousness" of our teleological reasoning—a conclusion that is also supported by evidence of some bias toward teleological explanations when forced to respond intuitively, and in some other situations (De Smedt and De Cruz 2020).

Aku Visala (2011, 178) has formulated the argument as follows based on Michael Murray's (2009, 169–171) analysis. Here HADD refers to the Hyperactive Agency Detection Device:

1. Most of the time HADD produces beliefs that we know to be false because it is hypersensitive.
2. If a mechanism produces many more false beliefs than true beliefs, then it is not reliable.
3. HADD is not reliable.
4. HADD is mostly responsible for our belief in God.
5. Therefore, belief in God is unjustified.

As Visala notes, premise 4 is questionable, as it remains far from proven that HADD is mostly responsible for religious belief. However, in addition to this critique, the most common response to the argument is to question premises 1 and 2, which ground the conclusion of unreliability. The point is that we rely on our agency detection device every day, and there is no reason to think that it is primarily unreliable, even though there are cases in which it fails (De Ridder 2014; van Eyghen 2019). Typically we are able to recognize quite quickly when we have erred in the detection of agency, such as when we mistake a garden hose for a snake. We can typically differentiate between a weak hunch that there is an agent, and a strong perception of qualities that seem to testify to the presence of a mind. The perception of purposefulness in biology is also far more enduring and widely held than examples of "promiscuous" teleological reasoning, such as explaining the pointiness of rocks as having the purpose that they "won't get smashed by animals" (Kelemen 1999, 1443). Moreover, as Barrett (2010) himself notes, the false positives created by HADD are related to danger, and theistic inferences are not made in similar circumstances. Thus, it seems that the unreliability of our agency detection should not be argued as a whole, but on a case-by-case basis, even taking the findings of the CSR into account. As Jong (2013, 529–530) points out, "We might freely assume that all supernatural agents that we detect are false positives, of course, but this would simply beg the question unless there is good reason for such an assumption. [...] Despite the rhetoric, most CSR theories argue that religious beliefs are products of properly functioning evolutionary adaptive and currently indispensable cognitive mechanisms."

Now it does seem that if the theories of the Cognitive Science of Religion (CSR) are correct, then there is a sense in which design-based explanations fit well with our cognitive architecture, and thus we may be predisposed to favor them at least in some situations. It is also clear from psychological studies and our common experience that human intuitions can be wrong. Nevertheless, the fact that some explanation fits with our cognitive architecture does not yet provide us with any reason to reject it. To insist that we should be immediately suspicious of any intuitively plausible views would be tantamount to saying that unintuitiveness is a merit of theories. In other words, we would believe these theories because they are absurd—"credo quia absurdum," as the early Christian theologian Tertullian (c. 155–240) is often wrongly understood as having stated (for the correction, see Vainio 2010, 24–31). Instead, as Visala (2011, 177) states, "there seems to be no reason whatsoever to think that the psychological plausibility of a belief should be taken as evidence for its falsity or irrationality."

Indeed, it seems to me that discarding the *prima facie* reliability of our commonsense reasoning would also lead to undercutting the possibility of scientific reasoning. This is because at least some compatibility between the human mind and the cosmos is required in order for the cosmos to be at all amenable to scientific discovery, and for human survival to have been possible in the first place. To reject intuitive, theological, and philosophical ideas about design just because they are not "science" according to one's definition would be a problematic form of scientism. However, in light of the possibility of errors, it would be beneficial to test our intuitions about design and see if they become stronger or weaker when all relevant considerations are taken into account—including evolution.

In the philosophical discussion on defeaters, two main types of defeaters have been proposed (Baker-Hytch and Benton 2015). First, there are *undercutting defeaters* that work by removing the reason to trust one's cognitive capacities in a particular case, rather than directly contradicting the belief. For example, the belief that I am wearing reality-augmenting VR goggles could be an undercutting defeater for the belief that I am presently perceiving a real, physical tree in front of me, since the defeater undercuts the reliability of my visual sense. There could still be a tree in front of me, but I would no longer have reason to think that I know about it. The second type, *rebutting defeaters*, work by providing evidence that directly contradicts my belief. For example, suppose that my perception of a tree is not that clear (perhaps I developed near-sightedness), and that someone whose perceptual capabilities I trust tells me that it is not in fact a tree, but a large bush. This would then function as a defeater for my belief, or at least weaken it greatly, since I would now have reason to think that my belief is not actually true.

In the case of design beliefs, two main undercutting defeaters have been proposed: (1) the results of the cognitive sciences of religion in making possible an evolutionary debunking argument against the rationality of religious belief, and (2) evolution in showing that the features of living organisms which trigger our design beliefs actually do not give evidence of design, but are the results of a natural process mimicking design. Evolution has also been proposed to be a rebutting defeater, but this would require evolution to directly contradict design beliefs. For example, if design beliefs were to require explaining the origin of species as a result of supernatural creation events, rather than natural processes, then evolution would contradict such beliefs. However, in view of the general compatibility of evolution and creation, it is more likely that evolution could succeed as an undercutting defeater.

Darwin himself stated that the "old argument of design in nature, as given by Paley, which formerly seemed to me so conclusive, fails, now that the law of natural selection has been discovered" (Darwin 1958 [1887], 87–88). Evolution is argued to produce "design without a designer," and thus show that even the strongest impressions of design are just an illusion. The alleged designedness of biological organisms would then be another example of how an intuitively rational, commonsensical understanding of the world has been refuted by a more detailed scientific examination. This is the central debunking argument that this book aims to respond to.

2.4 Conclusions

The idea that the order of nature is purposefully created is part of the doctrine of creation of the Abrahamic religions, which also finds support outside these religions, such as in strands of Hinduism. It is also an inference that is intuitively plausible for many based on the orderliness of nature, and many have claimed that this idea is rooted in the way the human mind instinctively interprets reality. However, in current discussion, many have claimed that the advancement of science makes belief in design unnecessary, or that the scientific evidence is even in contradiction with such belief. Here, two main points stand out as crucial for the salvaging operation.

First, in order to be theologically interesting, the design argument does not need to provide anything close to a proof of the Creator. Rather, the argument will already be valuable if it provides evidence that supports belief in creation, and a way to save the rationality of intuitive religious belief in design, at least within the framework of a Christian theology of nature. Moreover, though it can be important to consider the evidence of design in isolation, for the purposes of determining the strength of this particular piece of evidence, it is nevertheless clear that our ultimate conclusions about design will also be influenced by our total worldview, as well as other reasons and arguments. In order to save some value from the design argument, then, we do not have to demonstrate that design is a rationally necessary conclusion from the evidence—it just has to be a rationally permissible one, at least for those who do not think other reasons beyond evolution disqualify it.

Second, while this argument requires the rejection of scientism, it is nevertheless important to engage with the scientific results which have traditionally been held to conflict with the biological design argument. The rejection of scientism (as a philosophy about science) does not mean the rejection or disvaluing of the natural sciences as the empirical study of nature. In order to provide a meaningful alternative to the anti-evolutionists' defense of the design argument, the salvaging operation should not result in a "God of the gaps" argument, where divine action is posited as an explanation in an area where the evidence makes it more reasonable to expect a natural explanation. The argument should build on an understanding of divine providence which understands that God does not act merely (or even mostly) in gaps in natural processes.

As noted, though evolution and creation are broadly compatible, it is still in principle possible for evolution to function as an undercutting defeater that removes a reason for belief in creation. But before we understand evolutionary explanations and how they might have an impact on the design argument, we should first be clear about what the design argument is, and how its logic might be understood. I will turn to this task in the following chapter.

Notes

1. Research into the historical interaction of various forms of religion and science has shattered simplistic narratives of conflict, but also calls overly simplistic narratives of harmony into question (Brooke 1991; Harrison 2015).
2. Some have taken the issue of scientism to be a "canard," arguing that scientism is a strawman constructed by religious apologists (e.g., Coyne 2015, chapter 4). However, many such writers curiously go on to affirm scientistic sentiments elsewhere in the same books. For a review pointing this out regarding Coyne, see Feser (2016).
3. See de Ridder, Peels and van Woudenberg, René, eds. (2018); Stenmark (2001), and Peels (2017). A recent article disputing the critique of scientism is Hietanen et al. (2020). However, this article's defense of moderate scientism allows for the rationality of philosophical beliefs, which is all that is required for my own argument in this book, so I will not deal with the defense of moderate scientism further. For overall discussions of religious rationality in relation to science, see Stenmark (1995), Vainio (2010), and McGrath (2018).
4. Other prominent defenses of natural theology include Sennett and Groothuis ed. (2005), Sudduth (2009), and White (2016).
5. The most prominent anti-evidentialist in the contemporary discussion is probably D. Z. Phillips (1976), who argued that the evidentialists fundamentally misunderstand the nature of religious faith. See further Koistinen (2015). Alvin Plantinga is also sometimes presented as a fideist, but he allows for a positive role for natural theology. See Mascord (2007, chapter 6).
6. Quoted in Abraham (2017, 1).
7. For example, Meek (2011). Moser (2010) emphasizes the personal nature of religious knowledge to the extent of rejecting natural theology, arguing

that natural theology lacks the morally transformative nature of proper religious knowledge. For a response, see Dougherty and Rickabaught (2017), and for a counter-response Moser and Neptune (2017).

8. Pannenberg (1991) helpfully distinguishes between the concepts of *natural revelation* as the idea that God has embedded a revelation of himself in nature, and *natural theology* as the human attempt to reason based on this. When natural theology is understood to be built on natural revelation, the criticism that it bypasses the necessity of grace seems to miss the mark.
9. McGrath (2016, 18–22) argues for a "thick" definition of natural theology, where natural theology as a whole consists of many different and interrelating practices.
10. Note that Swinburne (2005, 119–120) also acknowledges that even if supernatural grace is required to accept the evidence of natural theology, it could still be natural.
11. Quoted in Wahlberg (2014, 53).
12. However, many defenders of design arguments would argue that the evidence of design in nature is stronger than in the ambiguous examples used in these studies.
13. As I have noted (Kojonen 2017b), this definition of methodological naturalism is not the only one in the literature. There has also been critique of methodological naturalism outside the ID movement (Torrance 2017). The historical roots and reasons for adopting methodological naturalism are explored further by Perry and Lane Ritchie (2018). My project fits comfortably within the kind of "science-engaged theology" that Perry and Ritchie defend as compatible with methodologically naturalistic science.
14. In contrast to this assumption, George (2016, 273–274) argues that Paley was moved to consider God also in other things, but simply thought that the orderly contrivance of biological organisms was the most convincing example to be used for the purposes of his argument.
15. A related objection is that the design argument would make God into a "cause among causes". For discussion of and responses to this objection, see Silva (2014) and Sollereder (2015).
16. As the editor of the ID-friendly open access journal *Bio-Complexity*, Leisola is one of the most influential supporters of ID outside the US.
17. The example of two people meeting by accident comes from Aquinas, *Summa Theologiae*, I 22.2 ad1. Paley (2008 [1802], XXVI) also uses a similar example.
18. However, many evolutionary biologists now argue that epigenetic processes do cause non-random inherited changes in response to the environment (Jablonka and Lamb 2010; Wagner 2012; Chan and Gordenin 2015). For further perspectives on chance in evolution see Ramsey and Pence (eds.) (2016).

19. Rosenberg (2017) disagrees, arguing that for mutations to be truly indeterministic, even God cannot predict them. Philipse (2013) also argues that this is not a plausible understanding of divine action. But for a further defense of compatibility, see van Woudenberg and Rothuizen-van der Steen (2015).
20. Dawkins' phrase is often quoted by proponents of ID. Loesberg (2007, 108–109) even argues that Dawkins should be recognized as one of the main thinkers behind the rise of the ID movement. I find this quite plausible.
21. This is not to deny that there have also historically been popular-level versions of design arguments based on the general orderliness of the cosmos and its fitness for life. An early version is provided by the Jewish philosopher Philo (20 BC–40 AD), who compared the cosmos to a well-built house (Philo, *Legum Allegoriae* III, 32, 98–99).
22. Wahlberg (2012) presents an in-depth defense of the perceptual nature of design beliefs, building on the work of philosopher of mind John McDowell (e.g., McDowell 1994).
23. In practice this would work much like understanding design beliefs as the products of intuitive or conscious inferences. At this point semiotics might help develop the argument further. See Southgate (2018); McGrath (2016, 94–98).

References

Abraham, William. 2017. Introduction. In *The Oxford Handbook of the Epistemology of Theology*, ed. William J. Abraham and Frederick D. Aquino. Oxford: Oxford University Press.

———. 2019. Particular Providence and the Details of Divine Action. In *Divine Action and Providence: Explorations in Constructive Dogmatics*, ed. Oliver D. Crisp and Fred Sanders, 40–54. Grand Rapids, MI: Zondervan Academic.

Abraham, William J., and Frederick D. Aquino, eds. 2017. *The Oxford Handbook of the Epistemology of Theology*. Oxford: Oxford University Press.

Alston, William. 1991. *Perceiving God: The Epistemology of Religious Experience*. Ithaca, NY: Cornell University Press.

Argyll, George. 1885. What Is Science? *Good Words*: 236–245.

Aristotle. 2000 [350 B.C.]. *Nicomachean Ethics*, Trans. Roger Crisp. Cambridge: Cambridge University Press.

Baker-Hytch, Max. 2018. Complexly Based Beliefs and the Generality Problem for Reliabilism. *Quaestiones Disputatae* 8 (2): 19–35.

Baker-Hytch, Max, and Matthew A. Benton. 2015. Defeatism Defeated. *Philosophical Perspectives* 29 (1): 40–66.

von Balthasar, Hans Urs. 1991. *The Glory of the Lord, Vol. V: The Realm of Metaphysics in the Modern Age*, Trans. O. Davies, A. Louth, B. McNeil, J. Saward, & R. Williams. San Francisco: Ignatius Press.

Barbour, Ian G. 1966. *Issues in Science and Religion*. New York, NY: Prentice Hall and Harper.

———. 1997. *Barbour Religion and Science: Historical and Contemporary Issues. A Revised and Expanded Edition*. San Francisco, CA: HarperOne.

Barrett, Justin. 2004. *Why Would Anyone Believe in God?* Walnut Creek: AltaMira Press.

———. 2010. The Relative Unnaturalness of Atheism: On Why Geertz and Markusson Are Both Right and Wrong. *Religion* 40 (3): 169–172.

———. 2012. *Born Believers: The Science of Childhood Religion Belief.* New York: Free Press.

Bloom, Paul. 2009. Religious Belief as an Evolutionary Accident. In *The Believing Primate: Scientific, Philosophical and Theological Reflections on the Origin of Religion*, ed. Michael Murray and Jeffrey Schloss, 118–127. New York, NY: Oxford University Press.

Brooke, John Hedley. 1991. *Science and Religion: Some Historical Perspectives.* Cambridge: Cambridge University Press.

———. 2002. Natural Theology. In *Science and Religion: A Historical Introduction*, ed. Gary Ferngren, 163–164. Baltimore, MA: The John Hopkins University Press.

Brooke, John Hedley, Fraser Watts, and Russell Re Manning, eds. 2013. *The Oxford Handbook of Natural Theology*. Oxford: Oxford University Press.

Brown, C. Mackenzie. 2008. The Design Argument in Classical Hindu Thought. *International Journal of Hindu Studies* 12 (2): 103–151.

Burns, Aaron. 2017. A Phenomenal Conservative Perspective on Religious Experience. *International Journal for Philosophy of Religion* 81 (3): 247–261. https://doi.org/10.1007/s11153-015-9557-7.

Burrell, David B., ed. 2010. *Creation and the God of Abraham.* Cambridge: Cambridge University Press.

Cavanaugh, William T. & Smith, James K. A. Eds. 2017. *Adam and the Fall.* Grand Rapids, MI: Eerdmans, 2017.

Chan, Kin, and Dmitry A. Gordenin. 2015. Clusters of Multiple Mutations: Incidence and Molecular Mechanisms. *Annual Review of Genetics* 49: 243–267.

Chignell, Andrew. 2018. The Ethics of Belief. *The Stanford Encyclopedia of Philosophy* (Spring 2018 Edition), Ed. Edward N. Zalta. Available at https://plato.stanford.edu/archives/spr2018/entries/ethics-belief/.

Clark, Kelly James, and Justin Barrett. 2011. Reidian Religious Epistemology and the Cognitive Science of Religion. *Journal of the American Academy of Religion* 79 (3): 639–675.

Coyne, Jerry. 2015. *Faith Versus Fact: Why Science and Religion Are Incompatible.* New York, NY: Viking Penguin.

Craig, William Lane, and Paul Copan. 2004. *Creation Out of Nothing: A Biblical, Philosophical, and Scientific Exploration.* Grand Rapids, MI: Baker Academic.

Craig, William Lane, and J.P. Moreland, eds. 2012. *The Blackwell Companion to Natural Theology.* Oxford: Blackwell Publishing.

Crick, Francis. 1989. *What Mad Pursuit: A Personal View of Scientific Discovery.* London: Weidenfeld and Nicolson.

Cunningham, Conor. 2010. *Darwin's Pious Idea: Why the Ultra-Darwinists and Creationists Both Get It Wrong.* Grand Rapids, MI: William B. Eerdmans.

Darwin, Francis. 1887. *The Life and Letters of Charles Darwin, Including an Autobiographical Chapter.* Vol. 1. London: John Murray.

Darwin, Charles. 1958 [1887]. In *The Autobiography of Charles Darwin*, ed. Nora Barlow. London: Collins.

Dawkins, Richard. 1986. *The Blind Watchmaker.* Harlow: Longman.

———. 2006. *The God Delusion.* London: Bantam Press.

De Cruz, Helen. 2014. The Enduring Appeal of Natural Theological Arguments. *Philosophy Compass* 9 (2): 145–153.

De Cruz, Helen, and Johan De Smedt. 2015. *A Natural History of Natural Theology: The Cognitive Science of Theology and Philosophy of Religion.* Cambridge, MA: The MIT Press.

De Ridder, Jeroen. 2014. Design Discourse and the Cognitive Science of Design. *Philosophia Reformata* 79 (1): 37–53.

De Ridder, Jeroen, Rik Peels, and René van Woudenberg. 2018. *Scientism: Prospects and Problems.* Oxford: Oxford University Press.

De Smedt, Johan, and Helen De Cruz. 2020. Is Intuitive Teleological Reasoning Promiscuous? In *Teleology and Modernity*, ed. William Gibson, Dan O'Brien, and Marius Turda, 185–202. Abingdon and New York, NY: Routledge.

Dembski, William A., ed. 1999. *Intelligent Design: The Bridge between Science and Theology.* Downers Grove, IL: InterVarsity Press.

Dougherty, Trent. 2018. Seemings, Reasons, and Knowledge: A Defense of Phenomenal Conservatism. In *Believing in Accordance with the Evidence: New Essays on Evidentialism*, ed. Kevin McCain, 39–51. Dordrecht: Springer.

Dougherty, Trent, and Brandon Rickabaugh. 2017. Natural Theology, Evidence and Epistemic Humility. *European Journal for Philosophy of Religion* 9 (2): 19–42.

Ellis, George F. 2018. Necessity, Purpose and Chance. In *God's Providence and Randomness in Nature: Scientific and Theological Perspectives*, ed. Robert J. Russell and Joshua M. Moritz, 21–67. West Conshohocken, PA: Templeton Press.

Esther Lightcap Meek. 2011. *Loving to Know: Covenant Epistemology.* Eugene, OR: Wipf & Stock.

Evans, C. Stephen. 2012. *Natural Signs and the Knowledge of God: A New Look at Theistic Arguments*. New York, NY & Oxford: Oxford University Press.

———. 2014. *God and Moral Obligation*. Oxford: Oxford University Press.

———. 2018. The Naïve Teleological Argument: An Argument from Design for Ordinary People. In *Two Dozen (Or So) Arguments for God's Existence*, ed. Jerry L. Walls and Trent Dougherty, 108–122. Oxford: Oxford University Press.

van Eyghen, Hans. 2019. Is Supernatural Belief Unreliably Formed? *International Journal for Philosophy of Religion* 85 (2): 125–148.

Fergusson, David. 2018. *The Providence of God: A Polyphonic Approach*. Cambridge: Cambridge University Press.

Feser, Edward. 2016, February. Omnibus of Fallacies. *First Things*. Available at https://www.firstthings.com/article/2016/02/omnibus-of-fallacies.

Futuyma, Douglas. 1983. *Science on Trial: The Case for Evolution*. New York, NY: Pantheon Books.

Ganeri, Martin. 2015. *Indian Thought and Western Theism: The Vedānta of Rāmānuja*. New York: NY & London.

Garvey, Jon. 2019. *God's Good Earth: The Case for an Unfallen Creation*. Eugene, OR: Cascade Books.

George, Marie. 2016. Thomistic Rebuttal of Some Common Objections to Paley's Argument from Design. *New Blackfriars* 97 (1069): 266–288.

Gettier, Edmund. 1963. Is Justified True Belief Knowledge? *Analysis* 23 (6): 121–123.

Giberson, Karl, ed. 2016. *Chance and Providence in the Monotheistic Traditions*. Oxford: Oxford University Press.

Hanby, Michael. 2016. *No God, No Science: Theology, Cosmology, Biology*. Oxford: Wiley-Blackwell.

Harrison, Peter. 2015. *The Territories of Science and Religion*. Chicago, IL: University of Chicago Press.

Haught, John F. 2004. Darwin, Design and Divine Providence. In *Debating Design: From Darwin to DNA*, ed. William A. Dembski and Michael Ruse, 229–244. Cambridge: Cambridge University Press.

Henriksen, Jan-Olav. 2015. A New Basis for Natural Religion? Recent Explanations of Religion and Their Challenges to Contemporary Philosophy of Religion. *Neue Zeitschrift für systematische Theologie und Religionsphilosophie* 57 (4): 464–482.

Hietanen, J., P. Turunen, I. Hirvonen, J. Karisto, I.O.T. Pättiniemi, and H. Saarinen. 2020. How Not to Criticise Scientism. *Metaphilosophy* 51 (4): 522–547.

Holder, Rodney. 2012. *The Heavens Declare: Natural Theology and the Legacy of Karl Barth*. West Conshohocken, PA: Templeton Press.

Hume, David. 1991. *An Enquiry Concerning Human Understanding*. South Bend, IN: Infomotions.

———. 2001 [1779]. *Dialogues Concerning Natural Religion*. South Bend, IN: Infomotions, Inc.

van Huyssteen, Wentzel. 1999. *The Shaping of Rationality: Toward Interdisciplinarity in Theology and Science*. Grand Rapids, MI: Eerdmans.

Jablonka, Eva, and Lamb, Marion J. 2010. Transgenerational Epigenetic Inheritance. In *Evolution: The Extended Synthesis*, ed. Massimo Pigliucci and Gerd B. Müller, 137–174. Cambridge, MA: The MIT Press.

Johnson, Phillip E. 1993. *Darwin on Trial*. 2nd ed. Downers Grove, IL: InterVarsity Press.

———. 1995. *Reason in the Balance*. Downers Grove, IL: InterVarsity Press.

———. 2007. Intelligent Design in Biology: the Current Situation and Future Prospects. *Think. The Royal Institute of Philosophy*, February 19. Available at https://www.discovery.org/a/3914/.

Jong, Jonathan. 2013. Explaining Religion (Away?). *Sophia* 52 (3): 521–533.

Juurikkala, Oskari. 2019. *The Patristic and Medieval Metaphor of the Book of Nature: Implications for Fundamental Theology*. Doctoral Dissertation, Pontificia Universita Della Santa Croce Facolta Di Teologia, Rome.

Kahane, Guy. 2011. Evolutionary Debunking Arguments. *Nous* 45 (1): 103–125.

Kant, Immanuel. 1960 [1793]. *Religion Within the Limits of Reason*. Trans. John Silber. New York, NY: Harper & Row.

Kelemen, Deborah. 1999. Why Are Rocks Pointy? Children's Preference for Teleological Explanations of the Natural World. *Developmental Psychology* 35 (6): 1440–1452.

———. 2004. Are Children 'Intuitive Theists'? Reasoning about Purpose and Design in Nature. *Psychological Science* 15 (5): 295–301.

Koistinen, Timo. 2015. D. Z. Phillips' Contemplative Philosophy and Theological Realism. *Philosophy, Culture, and Traditions: A Journal of the World Union of Catholic Philosophical Societies* 11: 43–56.

Kojonen, Erkki Vesa Rope. 2016a. *The Intelligent Design Debate and the Temptation of Scientism*. London: Routledge.

———. 2016b. The God of the Gaps, Natural Theology and Intelligent Design. *The Journal of Analytic Theology* 4. https://doi.org/10.12978/jat.2016-4.041708101413a.

———. 2017a. Natural Theology in Evolution: A Review of Critiques and Changes. *The European Journal of Philosophy of Religion* 9 (2): 83–117.

———. 2017b. Methodological Naturalism and the Truth Seeking Objection. *International Journal for Philosophy of Religion* 81 (3): 335–355.

———. 2020. Barth and the Return of Natural Theology. *Kerygma & Dogma* 66 (1): 41–67.

Kretzmann, Norman. 1999. *The Metaphysics of Creation: Aquinas's Natural Theology in Summa Contra Gentiles II*. Oxford: Oxford University Press.

Lamont, John. 2004. *Divine Faith*. Aldershot: Ashgate.

Launonen, Lari. 2018. The Naturalness of Religion: What It Means and Why It Matters. *Neue Zeitschrift für Systematische Theologie und Religionsphilosophie* 60 (1): 84–102.

Leech, David, and Aku Visala. 2011. The Cognitive Science of Religion: A Modified Theist Response. *Religious Studies* 47 (3): 301–316.

Leisola, Matti. 2017. Evolution: A Story Without a Mechanism. In *Theistic Evolution: A Scientific, Philosophical and Theological Critique*, ed. J.P. Moreland, Stephen C. Meyer, Christopher Shaw, Ann K. Gauger, and Wayne Grudem, 139–163. Wheaton, IL: Crossway.

Lenoir, Timothy. 1992. *The Strategy of Life*. Chicago, IL: University of Chicago Press.

Levering, Matthew. 2016a. Augustine on Creation: An Exercise in the Dialectical Retrieval of the Ancients. In *Wisdom and the Renewal of Catholic Theology: Essays in Honor of Matthew L. Lamb*, ed. Thomas P. Harmon and Roger W. Nutt, 49–65. Eugene, OR: Pickwick.

———. 2016b. *Proofs of God: Classical Arguments from Tertullian to Barth*. Grand Rapids, MI: Baker Academic.

———. 2017. *Engaging the Doctrine of Creation: Cosmos, Creatures, and the Wise and Good Creator*. Grand Rapids, MI: Baker Academic.

Loesberg, Jonathan. 2007. Kant, Hume, Darwin, and Design: Why Intelligent Design Wasn't Science Before Darwin and Still Isn't. *The Philosophical Forum* 38 (2): 95–123.

Loke, Andrew. 2017. *God and Ultimate Origins: A Novel Cosmological Argument*. Basingstoke: Palgrave.

Loke, Andrew Ter Ern. 2019. Theological Critiques of Natural Theology: A Reply to Andrew Moore. *Neue Zeitschrift Für Systematische Theologie und Religionsphilosophie* 61 (2): 207–222.

Luther, Martin. 2016 [1526]. *Der Grosse Catechism*. Hamburg: Severus.

MacIntyre, Alister. 2016. *Whose Justice? Which Rationality?* Notre Dame, IN: University of Notre Dame Press. Original publication 1988.

MacSwain, Robert. 2013. *Solved by Sacrifice: Austin Farrer, Fideism and the Evidence of Faith*. Leuven: Peeters.

Mascord, Keith. 2007. *Alvin Plantinga and Christian Apologetics*. Eugene, OR: Wipf & Stock.

McCord Adams, Marilyn. 2014. What's Wrong with the Ontotheological Error? *Journal of Analytic Theology* 2. https://doi.org/10.12978/jat.2014-1.120013000318a.

McDowell, John. 1994. *Mind and World*. Cambridge, MA: Harvard University Press.

McGrath, Alister. 2001. *A Scientific Theology, Vol. 1: Nature*. Edinburgh: T & T Clark.

———. 2004. *The Science of God: An Introduction to Scientific Theology*. London: T&T Clark International.

———. 2008. *The Open Secret: A New Vision for Natural Theology*. Oxford: Blackwell Publishing.

———. 2009. *Darwinism and the Divine: Evolutionary Thought and Natural Theology*. Oxford: Wiley-Blackwell.

———. 2016. *Re-Imagining Nature: The Promise of a Christian Natural Theology*. Oxford: Wiley-Blackwell.

———. 2018. *The Territories of Human Reason: Science and Theology in an Age of Multiple Rationalities*. Oxford: Oxford University Press.

McLeish, Tom. 2014. *Faith and Wisdom in Science*. Oxford: Oxford University Press.

McMullin, Ernan. 2009. On the Origin of Terrestrial Life: A Christian Perspective. In *Exploring the Origin, Extent, and Future of Life: Philosophical, Ethical, and Theological Perspectives*, ed. Constance M. Bertka, 80–95. Cambridge: Cambridge University Press.

Moon, Andrew. 2016. Recent Work in Reformed Epistemology. *Philosophy Compass* 11 (12): 879–891.

Moreland, J.P., ed. 1994. *The Creation Hypothesis: Scientific Evidence for an Intelligent Designer*. Downers Grove, IL: IVP Books.

Moser, Paul K. 2010. *The Evidence for God: Religious Knowledge Re-examined*. New York, NY: Cambridge University Press.

Moser, Paul K., and Clinton Neptune. 2017. Is Traditional Natural Theology Cognitively Presumptuous? *The European Journal for Philosophy of Religion* 9 (2): 213–222.

Mullen, John T. 2004. *Design Arguments within a "Reidian" Epistemology*. Unpublished Dissertation, University of Notre Dame, Notre Dame, IN.

Murray, Michael. 2009. Scientific Explanation of Religion and the Justification of Religious Belief. In *The Believing Primate: Scientific, Philosophical and Religious Reflections on the Origin of Religion*, ed. Jeffrey Schloss and Michael Murray, 168–178. Oxford: Oxford University.

Newman, John Henry. 1974 [1870]. Letter to William Robert Brownlow, April 13, 1870. In *Newman, John Henry, the Letters and Diaries of John Henry Newman*, ed. Charles Stephen Dessain and Thomas Gornall, vol. 25. Oxford: Clarendon Press.

Paley, William. 2008 [1802]. *Natural Theology*. Oxford World's Classics. Oxford: Oxford University.

Pannenberg, Wolfhart. 1991. *Systematic Theology*. Vol. 1. Edinburgh: T & T Clark.

Peels, Rik. 2017. The Fundamental Argument against Scientism. In *Science Unlimited? The Challenges of Scientism*, ed. M. Boudry and M. Pigliucci, 165–184. Chicago, IL: Chicago University Press.

Peels, Rik, Jeroen de Ridder, and René van Woudenberg, eds. 2020. *Scientific Challenges to Common Sense Philosophy*. London: Routledge.

Perry, John, and Sarah Lane Ritchie. 2018. Magnets, Magic, and Other Anomalies: In Defense of Methodological Naturalism. *Zygon: Journal of Religion and Science*. 53 (4): 1064–1093.

Peters, Ted, and Martinez Hewlett. 2003. *Evolution from Creation to New Creation: Conflict, Conversation and Convergence*. Nashville, TN: Abingdon Press.

Philipse, Herman. 2013. The Real Conflict between Science and Religion: Alvin Plantinga's Ignoratio Elenchi. *European Journal for Philosophy of Religion* 5: 87–110.

Phillips, D.Z. 1976. *Religion Without Explanation*. Oxford: Blackwell.

Plantinga, Alvin. 2011. *Where the Conflict Really Lies: Science, Religion, and Naturalism*. New York, NY: Oxford University Press.

Pollock, John L. 1986. *Contemporary Theories of Knowledge*. Totowa, NJ: Rowman and Littlefield.

Pruss, Alexander, and Joshua L. Rasmussen. 2018. *Necessary Existence*, 2018. Oxford: Oxford University Press.

Ramsey, William. 2019. Eliminative Materialism. *The Stanford Encyclopedia of Philosophy* (Spring 2019 Edition), Ed. Edward N. Zalta. Available at https://plato.stanford.edu/archives/spr2019/entries/materialism-eliminative/.

Ramsey, Grant H., and Charles H. Pence. 2016. *Chance in Evolution*. Chicago, IL: University of Chicago Press.

Ratzsch, Del. 2003. Perceiving Design. In *God and Design: The Teleological Argument and Modern Science*, ed. Neil A. Manson, 125–145. London: Routledge.

———. 2009. Humanness in Their Hearts: Where Science and Religion Fuse. In *The Believing Primate: Scientific, Philosophical and Religious Reflections on the Origin of Religion*, ed. Jeffrey Schloss and Michael Murray, 215–245. Oxford: Oxford University.

———. 2010. Teleological Arguments for God's Existence. *Stanford Encyclopedia of Philosophy*. Winter 2010 Edition, Ed. Edward N. Zalta. Available at http://plato.stanford.edu/archives/win2010/entries/teleological-arguments/.

Reid, Thomas. 2002 [1785]. In *Essays on the Intellectual Powers of Man (1785)*, ed. Derek R. Brookes. University Park, PA: Penn State Press.

Ridder, De, and Jeroen. 2014. Design Discourse and the Cognitive Science of Design. *Philosophia Reformata* 79 (1): 37–53.

Riskin, Jessica. 2016. *The Restless Clock: A History of the Centuries-Long Argument Over What Makes Living Things Tick*. Chicago: The University of Chicago Press.

Rosenberg, Alex. 2011. *The Atheist's Guide to Reality: Enjoying Life Without Illusions*. New York, NY: W. W. Norton & Co.

———. 2017. "Theism and Allism." *Being, Freedom, and Method: Themes from the Philosophy of Peter van Inwagen*. Ed. John A. Keller. Oxford: Oxford University Press, 227–240.

Rosenberg, Stanley, Michael Burdett, Michael Lloyd, and Benno van den Toren, eds. 2018. *Finding Ourselves After Darwin: Conversations on the Image of God, Original Sin, and the Problem of Evil.* Grand Rapids, MI: Baker Academic.

Ruse, Michael. 2001. *Can a Darwinian Be a Christian? The Relationship Between Science and Religion.* Cambridge: Cambridge University Press.

———. 2018. *Darwinism as Religion: What Literature Tells Us about Evolution.* Oxford: Oxford University Press.

Russell, Robert J. 2019. What We've Learned from Quantum Mechanics About Noninterventionist Objective Divine Action in Nature – and Its Remaining Challenges. In *God's Providence and Randomness in Nature*, ed. Robert J. Russell and Joshua M. Moritz, 133–172. West Conshohocken, PA: Templeton Press.

Sedley, David. 2007. *Creationism and Its Critics in Antiquity.* Sather Classical Lectures 66. Berkeley, CA: University of California Press.

Sennett, James F., and Douglas Groothuis, eds. 2005. *In Defence of Natural Theology: A Post-Humean Assessment.* Downers Grove, IL: InterVarsity Press.

Shaw, Kegan. 2019. A Plea for the Theist in the Street: A Defense of Liberalism in the Epistemology of Religious Experience. *Faith and Philosophy* 36 (1): 102–128.

Silva, Ignacio. 2014. Great Minds Think (Almost) Alike: Thomas Aquinas and Alvin Plantinga on Divine Action in Nature. *Philosophia Reformata* 79 (1): 8–20.

Sober, Elliott. 2011. Evolution Without Naturalism. In *Oxford Studies in Philosophy of Religion*, ed. Jonathan Kvanvig, vol. 3, 187–221. Oxford: Oxford University Press.

Sollereder, Bethany. 2015. A Modest Objection: Neo-Thomism and God as a Cause among Causes. *Theology and Science* 13 (3): 1–9.

———. 2018. *God, Evolution, and Animal Suffering: Theodicy Without a Fall.* London: Routledge.

Southgate, Christopher. 2008. *The Groaning of Creation: God, Evolution, and the Problem of Evil.* Louisville, KY: Westminster John Knox Press.

———. 2018. *Theology in a Suffering World: Glory and Longing.* Cambridge: Cambridge University Press.

Stenmark, Mikael. 1995. *Rationality in Science, Religion and Everyday Life: A Critical Evaluation of Four Models of Rationality.* Notre Dame, IN: University of Notre Dame Press.

———. 2001. *Scientism: Science, Ethics and Religion.* Burlington: Ashgate Publishing Co.

Stich, Stephen P. 1985. Could Man Be an Irrational Animal? Some Notes on the Epistemology of Rationality. *Synthese* 64 (1): 115–135.

Stump, Eleonore. 2016. *The God of the Bible and the God of the Philosophers.* The Aquinas Lectures. Vol. 80. Milwaukee, WI: Marquette University.

Sudduth, Michael. 2009. *The Reformed Objection to Natural Theology*. Farnham: Ashgate.

Swanke, Johannes. 2014. Luther's Theology of Creation. In *The Oxford Handbook of Martin Luther's Theology*, ed. Robert Kolb, Irene Dingel, and Batka Lubomir, 201–211. Oxford: Oxford University Press.

Swinburne, Richard. 2003. *The Resurrection of God Incarnate*. Oxford: Oxford University Press.

———. 2004a. *The Existence of God*. Oxford: Oxford University Press.

———. 2004b. Natural Theology, its 'Dwindling Probabilities' and 'Lack of Rapport'. *Faith and Philosophy* 21 (4): 533–546.

———. 2005. *Faith and Reason*. 2nd ed. Oxford: Oxford University Press.

———. 2011. Evidence. In *Evidentialism and Its Discontents*, ed. Trent Dougherty. Oxford: Oxford University Press.

———. 2012. *Mind, Brain & Free Will*. Oxford: Oxford University Press.

Torrance, Andrew. 2017. Should a Christian Adopt Methodological Naturalism? *Zygon* 52 (3): 691–725.

Torrance, Andrew, and Thomas McCall, eds. 2018a. *Christ and the Created Order: Perspectives from Theology, Philosophy, and Science*. Grand Rapids, MI: Zondervan Academic.

———, eds. 2018b. *Knowing Creation: Perspectives from Theology, Philosophy and Science*. New York, NY: Zondervan Academic.

Tracy, Thomas F. 2013. Divine Purpose and Evolutionary Processes. *Zygon* 48 (2): 454–465.

Turner, Denys. 2004. *Faith, Reason and the Existence of God*. Cambridge: Cambridge University Press.

Ulanowitz, Robert. 2016. Process Ecology: Making Room for Creation. *Sophia* 55: 357–380.

Vainio, Olli-Pekka. 2010. *Beyond Fideism: Negotiable Religious Identities*. Burlington, VT: Ashgate.

———. 2020. *Religious Language*. Cambridge Elements. Cambridge: Cambridge University Press.

Van Woudenberg, René, and J. Rothuizen-van der Steen. 2015. Both Random and Guided. *Ratio* 28 (3): 332–348.

Vatican International Theological Commission. 2004. *Communion and Stewardship: Human Persons Created in the Image of God*. Available at http://www.vatican.va/roman_curia/congregations/cfaith/cti_documents/rc_con_cfaith_doc_20040723_communion-stewardship_en.html.

Visala, Aku. 2011. *Theism and the Cognitive Science of Religion*. Farnham: Ashgate.

von Wachter, Daniel. 2014. Defending Design Arguments Against Plantinga. *Philosophia Reformata* 79 (1): 54–65.

Wagner, Andreas. 2012. The Role of Randomness in Darwinian Evolution. *Philosophy of Science* 79 (1): 95–119.

Wahlberg, Mats. 2012. *Reshaping Natural Theology: Seeing Nature as Creation*. London: Palgrave Macmillan.

———. 2014. *Revelation as Testimony: A Philosophical-Theological Study*. Grand Rapids, MI: Eerdmans.

Whewell, William. 1834. *Astronomy and General Physics: Considered with Reference to Natural Theology*. London: William Pickering.

White, Thomas Joseph, and O. P. 2016. *Wisdom in the Face of Modernity: A Study in Thomistic Natural Theology*. 2nd ed. Ave Maria, FL: Sapientia Press.

Wilkins, J.S., and P.E. Griffiths. 2013. Evolutionary Debunking Arguments in Three Domains: Fact, Value, and Religion. In *A New Science of Religion*, ed. G. Dawes and J. Maclaurin. London: Routledge.

Williams, Thomas. 2005. The Doctrine of Univocity is True and Salutary. *Modern Theology* 21 (4): 575–585.

CHAPTER 3

From Teleological Arguments to Thought Experiments

If design arguments were clearly irrational on the philosophical level, it would arguably not make much sense to inquire whether they are compatible with evolution. And indeed many seemingly weighty philosophical criticisms of design arguments have been presented over the years. Could these already rule out the possibility of design arguments a priori, before the consideration of the evidence? In this chapter, I will not attempt to analyze or respond to all philosophical criticisms—but I will comment on some of the prominent ones, and argue that the search for an alternative explanation for the orderliness of the cosmos is indeed important for criticizing design arguments.

Critics of biological design arguments also have great differences of opinion over what, exactly, might be wrong in these arguments (Ruse 2004; Sober 2003). Most prominently, there has been much disagreement on whether Hume's philosophical criticisms were sufficient to refute the design argument, or whether evolutionary theory was the crucial factor. Those who think evolutionary explanations were necessary feel that the evidence and logic used in the design argument is sound, and rationally compelling before Darwin. Dawkins counts himself among those emphasizing the importance of Darwin:

> I feel more in common with the Reverend William Paley than I do with the distinguished modern philosopher, a well-known atheist, with whom I once

E. V. R. Kojonen, *The Compatibility of Evolution and Design*, Palgrave Frontiers in Philosophy of Religion,
https://doi.org/10.1007/978-3-030-69683-2_3

> discussed the matter at dinner. I said that I could not imagine being an atheist at any time before 1859, when Darwin's *Origin of Species* was published. 'What about Hume?' replied the philosopher. 'How did Hume explain the organized complexity of the living world?', I asked. 'He didn't', said the philosopher. 'Why does it need any special explanation?'
>
> Paley knew that it needed a special explanation, Darwin knew it, and I suspect in his heart of hearts my philosopher companion knew it too. (Dawkins 1986, 5–6)

However, in practice Dawkins does also appeal to Humean philosophical considerations to justify his rejection of the design argument. Later, in *The Blind Watchmaker*, he also criticizes the explanatory power of design arguments in the case of the origin of life, arguing that "to explain the origin of the DNA/protein machine by invoking a supernatural Designer is to explain precisely nothing, for it leaves unexplained the origin of the Designer. You have to say something like 'God was always there', and if you allow yourself that kind of lazy way out, you might as well just say 'DNA was always there', or 'Life was always there', and be done with it" (Dawkins 1986, 141). A similar critique was already expressed by Hume in his *Dialogues*. For Hume, if we can accept the mind of a designer as the proper stopping point in our search for further explanations, we could just as well also accept the material world as the stopping point, and so avoid the conclusion of design (Hume 2001 [1779]; Loesberg 2007). Here it is clear that Dawkins also has strong metaphysical and philosophical reasons for his position. Many critics have even claimed that Dawkins' adherence to these principles is so strong that they make evolution the only theory that could even in principle provide a good explanation of biological complexity (Orr 2007; Plantinga 2007; Feser 2017, 249–168).

While I am critical of Dawkins' overall conclusion, in this chapter I will argue with him against those who hold that Humean critiques of biological design arguments are sufficient. It seems to me that design can indeed be really explanatory, and the philosophical objections can be bypassed or responded to. Even if not all readers will be convinced, at least this chapter will also lay the groundwork for considering how biological design arguments might be compatible or incompatible with mainstream evolutionary biology, and how they are linked with fine-tuning design arguments.

3.1 Explaining Evidence of Design

Personal Explanation

The debate over Intelligent Design shows some tendency to regard scientific mechanistic explanations as the model against which all other explanations should be evaluated (Kojonen 2016, 82–86). As noted in Chap. 2, the success of the natural sciences also contributes to a "scientism" in which the natural sciences are taken as always providing the best or even the only way of investigating the cosmos. Particularly in the old deductive-nomological model of explanation, it was argued that scientific explanations should always be based on the laws and regularities of nature, which are then used to deductively predict observations. The model has since come under critique, as it does not appear to account for all examples of scientific explanation (Woodward 2017).

However, as I argued in Chap. 2, rationality is not restricted to science. In the philosophy of religion, it has thus been common to point out that outside the traditional domain of the natural sciences, there is another type of explanation that all humans are familiar with: personal explanation by reference to the purposes, beliefs, intentions, and capabilities of agents. The distinction between personal explanation and scientific explanation has been defended in depth by Swinburne in the context of theistic explanation (Swinburne 2010, 2004a; for critique see Dawes 2009; Philipse 2012).[1] As noted, some in the contemporary discussion would dismiss the explanatory power of personal explanations altogether. But just because a type of explanation does not fit the criteria of explanation used in the natural sciences, this does not mean that the use of such explanations could not be reasonable or even necessary. This is similar to how the usefulness of metal detectors in finding metal does not show that no non-metallic objects exist (Feser 2017, 273–285). Because rationality and rational explanation are not restricted to the natural sciences, the core question is whether the design argument is a good argument, not whether it is scientific or not (Kojonen 2017).

Leaving aside the difficult problem of defining science, the explanatory power of personal explanations seems to be required for our normal everyday lives.[2] We explain people's actions by reference to their intentions, striving to understand reasons in order to make the complexity of human behavior intelligible, and although our explanations are fallible, these are far superior to just assuming that people act randomly. In the case of artifacts, we routinely distinguish between (seemingly) unintentional artifacts of human civilization, like garbage and tracks in the mud, and intentional artifacts, like the order of an automobile's parts. Similarly, in theistic design arguments it is claimed that nature has certain features because its order has been intentionally designed, and that these features give evidence of their purposive origin. As Ratzsch and Koperski (2019) state, design arguments are about "finding and identifying various traces of the operation of a mind in nature's temporal and physical structures, behaviors and paths."

Just as there are differences between scientific explanations and personal explanations, as traditionally understood, so too are there differences in evaluating them. For example, consider the criterion of testability as it is used in the natural sciences. Within the sciences, testing ideas against empirical results is rightly valued, since this testing allows us to find out whether the theory meshes with reality. Theories are tested not simply against current scientific knowledge, but also against future observations that are predicted by the model. For example, knowing the laws of thermodynamics, we can predict that the amount of usable energy in any given closed system will decrease over time. But deriving predictions can often be difficult, and requires the addition of auxiliary hypotheses to the theory. Moreover, in contemporary philosophy of science, the strong linking of predictive and explanatory capacity has come under question. It may be that scientific explanation is more concerned with understanding the causal structure of the universe, and explanatory unification, rather than being primarily concerned with prediction (Salmon 1990; Woodward 2003, 2017; Cleland 2011; McGrath 2019).

Personal explanations are also testable to some degree. For example, we can test our ideas about the goals of an agent against our observations of what that agent actually does, and we can test our ideas about the functioning of artificial systems by testing what each part does in practice. However, recall the thesis that our knowledge of an object should be adapted to its nature. As Ratzsch (2001, 117) notes, "design theories are, ultimately, theories involving agency. And, with respect to agents, theories and explanations often must trail data. We are often in positions of utter inability to predict specific human actions, but such actions once observed may be readily and quite legitimately explainable." It is difficult to predict the behavior of agents, even though we can still distinguish between purposeful and non-purposeful behavior, and note when something created by humans displays marks of design. We appear to have an innate capacity for doing this, and learn more both socially and through our own experience of designing and creating things. Knowledge of a particular agent's personality and motives may allow for more detailed predictions of that agent's actions, but we do not hold the ability to make predictions as a criterion of personal explanations, at least in the same exact way as in explanations involving natural phenomena. With the updated understanding of the nature of explanation in the philosophy of science, requiring exact predictions from design-based explanations seems unnecessary (Koons 2017), although the idea of design has historically had some heuristic value even within the sciences (Ratzsch 2001, chapter 11).

In some design arguments, the postulated motive and character of the designer play a part in the explanation. In others, it is the unique capacities that are linked with intentional agents that play the central role. Thus, Swinburne (2004a, 346–349, 2004b) argues that theism explains the orderliness and fine-tuning of the universe, because a good God would create bodily creatures capable of interacting with each other. Collins (2009, 254–256) similarly argues that our moral knowledge should lead us to think that it is plausible that a good God would create a cosmos allowing for the existence of complex life. Life is valuable, and thus a morally good God would value life. In this way, the moral goodness of God grants explanatory hypotheses involving such a God explanatory power. Such a God would value some worlds above others, and this narrows down what God would be expected to do. This then allows us to argue that the character of our cosmos fits or does not fit with this understanding of God.

By contrast, other design arguments leave the designer's identity and motives mysterious, and in them the moral character of the designer does not play a major role in providing explanatory power. For instance, the ID movement argues that certain types of patterns are known to be well explained by the properties possessed by intelligent causes, even if the motives of the intelligent designer are unknown. Similarly, McGrew (2004) argues that human experience can already provide us with inductive data that links designers to certain kinds of order, and that this already provides the design argument with sufficient explanatory power. Indeed, explanations referring to intentionality typically do not include the kind of mechanistic and mathematical detail that scientific explanations tend to provide (Collins 2006). For theistic explanations, no such mechanism could even in principle be given for everything, since according to theism God can bring about any result he chooses without any need for intermediate second causes. Thus requiring such a mechanism for explanatory power would already rule out theism based on epistemic criteria. But the problem of not being able to provide further details about the mechanism is not necessarily unique to theism: as Dawes (2009, 51–53) notes, in all explanations there comes a point where we reach the level of basic causal powers, and are unable to specify further intermediate mechanisms. To insist on an explanation at such a truly basic level would just lead to an infinite causal regression.

As critics of scientism have noted, it would also be self-defeating to claim that science is the ultimate criterion of what can count as rational belief, since such a philosophical criterion would not itself be the result of science. Instead, the way to falsify a philosophical or metaphysical argument is to criticize the premises of that argument, rather than trying to falsify the conclusion directly or claiming that the argument should contain more mathematics. For example, suppose that we argue that (following the *Kalam* cosmological argument, Loke 2017) (1) whatever begins to exist has a cause outside itself; (2) the universe began to exist; and (3) the universe therefore has a cause outside itself. In this case it would not be a refutation of the argument to say that we cannot falsify the existence of the postulated cause directly through some other experiment. Rather, if the premises of the argument are correct, then the conclusion is rationally undeniable, since it follows as a consequence of the premises. Alternatively, if the premises are only very probable rather than certain, then the conclusion will also only be very probable, depending on the probability of the premises. To deny this would be to deny the basic principles of logic without which all scientific arguments would also be impossible. Hence, to reject philosophical design arguments one needs to criticize their premises, not merely quibble about whether they fulfill some criterion of science.

Since Hume and Kant, explanation by reference to divine action has often been also criticized philosophically as being in principle impossible because of the limits of human experience (Swinburne 2012; Taliaferro 2013). Even though such considerations have not succeeded in making theistic explanations impossible, some still argue that they at least provide a reason to give theistic explanations a very low background probability. Dawes (2009, chapter 7), for example, argues that because our experience is only of bodily agents, the theistic concept of an immaterial Creator has low background probability. However, Swinburne (2012) argues that the concept of an immaterial Creator can be easily formulated based on our experience of causation and human persons. Even if we assume that this concept is so foreign that it could only have come to us through revelation, it can still be understood, and then we can ask what kind of understanding of God is best supported by the evidence. Nevertheless, the worry about our inability to know about the designer's motivations, and thus to predict a designer's actions, continually resurfaces in current discussions of design arguments. The worry, as De Cruz and De Smedt (2015, 74) put it, is that in "the case of divine action [...] we do not have empirical knowledge to draw upon, and thus no assumptions can be made about what God would or would not do." Discussion of these issues will quickly lead us deep into metaphysics and epistemology, but later in this chapter I will argue that thought experiments allow us to bypass at least the most extreme versions of this in principle-objection against design arguments.

The Fine-Tuning Design Argument

What kind of features do design arguments commonly appeal to? As noted, the fine-tuning design argument is based on the observation that the laws, constants, and starting conditions of the cosmos allow for the existence of complex life. It appears that the requirements for the emergence of complex life are very stringent, requiring the right kind of forces to exist in the right portions (Geraint and Barnes 2016). While there is some amount of leeway, in many cases even very small changes would make the existence of life as we know it (or even the existence of stable elements) impossible. Four main types of evidence can be recognized (Collins 2009). First (1), the suitable types of laws and forces need to exist. If, for example, one of the four basic forces of physics were missing, then fine-tuning the rest would not suffice. Forces that interfere with these to prevent life should also not exist. Second (2), the strengths of the basic forces in relation to each other must be suitable for life. Third (3), the matter and anti-matter of the early cosmos must have the correct type of properties and proportion. Fourth (4), the result of these factors, then, should be the type of elements that are suitable to be "building blocks" for life. In addition to these factors, many other considerations important for life and scientific discovery have been cited, such as the features of the cosmos that allow for scientific discovery (Collins 2018). In contrast to biological design arguments, however, "local" phenomena like the fine-tunedness of our solar system are not typically cited in fine-tuning arguments (for such an approach, see Gonzales and Richards 2004). Instead, fine-tuning design arguments focus on "global" phenomena that would affect any planet suitable for life.

Only a few critics of the argument (e.g., Stenger 2011) have questioned the existence of fine-tuning, as the data is acknowledged by scientists with widely different worldviews (Barnes 2013). However, the evaluation of the probability of each variable has proven to be a matter of dispute, as has the explanation of fine-tuning. Usually, to evaluate statistical probabilities, we need knowledge of the natural processes at work in the situation to be evaluated. For example, in the case of a six-sided die, we know that there are six possible positions that the die could have landed on. However, in the case of fine-tuning, we cannot scientifically determine the processes responsible, nor would this even be possible if the laws, constants, and starting conditions of the cosmos are directly created by God from nothing. Thus, the probability cannot be calculated statistically. This has been called the "normalization problem" (McGrew et al. 2003; Halvorson 2018). However, proponents of the fine-tuning design argument, such as Collins (2009, 226–252) and Barnes (2018), have countered that the point of the fine-tuning probability calculations is not to calculate the physical, statistical probability of the life-permitting values. Rather, the point is to calculate a reasonable epistemic probability for the proposition that the laws of the cosmos fall into the range that permits the existence of complex life. Some reasonable comparison range must be selected as a null hypothesis, and here proponents of fine-tuning arguments are simply following normal procedures. Moreover, the assigned probabilities can typically be justified with further arguments. For example, the extreme precision of the cosmological constant is set by the need to match the calculated total dark energy of the cosmos (Barnes 2013).

The other major response to fine-tuning has been to present alternative explanations. At the most basic level, it has been argued that our existence itself makes it impossible for us to observe any kind of universe that does not permit life. The anthropic principle, as popularized by Barrow and Tipler (1986, 16), states that "the observed values of all physical and cosmological quantities are not equally probable but they take on values restricted by the requirement that there exist sites where carbon-based life can evolve and by the requirement that the Universe be old enough for it to have already done so." Thus it can be argued that there is nothing surprising about us observing such a universe, because otherwise we could not even exist to observe anything. However, this does not yet provide an explanation for why and how the laws of nature are life-permitting and allow us to exist. Just as survival from a firing squad does not explain itself (but calls for an explanation), so also the fact that our existence is only possible because certain conditions are fulfilled does not yet explain why and how these conditions obtain (Leslie 1989, 13–15).

To better explain fine-tuning, the anthropic principle needs to be complemented with a multiverse hypothesis. Suppose for the sake of argument that there is an infinite or at least enormously large number of universes, with varying constants of nature and natural laws. In that case, there will be innumerable universes that are hostile to life, but any living observers will nevertheless always find themselves in the minute portion of universes which support life. After all, living creatures can only exist in these universes. Thus, if our part of the multiverse appears purposefully designed, then perhaps this is due to this anthropic selection effect, rather than due to design.

Proponents of the multiverse explanations agree with proponents of the fine-tuning design argument that something about the laws and constants of nature at least stands in need of explanation. They simply disagree about the proper explanation. As Susskind (2006, 343) puts it: “Our own universe is an extraordinary place that appears to be fantastically well designed for our own existence. This specialness is not something that we can attribute to lucky accidents, which is far too unlikely. The apparent coincidences cry out for an explanation.” However, he then goes on to argue that design is an unsatisfactory explanation, since it leaves many questions unanswered, such as “who designed the designer, by what mechanism the designer intervenes to guide evolution, whether the designer violates the Laws of Physics to accomplish its goals, and whether the designer is subject to the laws of quantum mechanics” (ibid.). These types of questions are often presented as unanswerable criticisms of design arguments, bypassing centuries of theology and metaphysics.

However, defenders of the fine-tuning design argument have questioned whether the naturalistic explanations truly explain the same thing as theistic design. As Collins (2009, 256–262) notes, naturalistic explanations of fine-tuning typically rely on finding a fundamental natural process that generates the fine-tuned universe either directly or by varying the laws and constants of universes, akin to a natural search engine. But this only appears to move the problem of fine-tuning to this new area. According to Collins, all postulated natural mechanisms to generate a multiverse would themselves need to be fine-tuned, in order to be able to produce a life-permitting universe even as one among others, and to enable the universe-generating and constant-varying process. Thus it simply seems to move the problem back. If searching for an explanation for fine-tuning is reasonable in the case of our cosmos, it will then also be reasonable to search for an explanation for the fine-tuning at this more fundamental level.[3]

In the multiverse hypothesis, the anthropic principle explains why we would find ourselves in a habitable cosmos. If we were not in a habitable cosmos, then we could not be observing at all; therefore, it is inevitable that any observers will find themselves in such conditions. However, at most, this only predicts our observation of the amount of fine-tuning that is absolutely necessary for our existence. But arguably, the fine-tuning that we can observe in our cosmos extends well beyond what is needed for survival. It also includes the rationality and intelligibility of the cosmos, which make it open to scientific discovery. To respond to this, proponents of the multiverse hypothesis can attempt to argue for the mediocrity of our cosmos. This has been a traditional critique of design arguments. Hume (2001 [1779], X, XI), for example, compared our cosmos to a poorly built house, and thus claimed that our observation of the cosmos could at most prove the existence of a bumbling, inept deity. Following Hume's lead, the proponent of the multiverse could argue that if the amount of fine-tuning can vary, it might be likely that we should find ourselves between the extremes of a supremely finely tuned cosmos and a cosmos containing the bare minimum. This does not seem like a sufficient response, given the great extent of fine-tuning beyond what is necessary. But the moral evaluation of the cosmos would quickly take us deep into theology and the philosophical discussion of theodicy.

Ultimately, preference for the explanation of design over naturalism, or vice versa, is also affected by the consonance of this explanation with our broader worldview. Proponents of design arguments who accept Christian theism thus commonly argue that the theistic hypothesis is less ad hoc than the multiverse hypothesis, since it was formulated and strongly believed already prior to the discovery of fine-tuning (Collins 2009, 205–209). The debate over the relative simplicity and background probabilities of theism and naturalism is also common in the discussion of such arguments. For committed naturalists who hold that they have other reasons to be skeptical of theism and to regard design-based explanations as absurd, there can be a strong presupposition against design as the explanation for fine-tuning. In that case the multiverse hypothesis (or another explanation, or simply refusing to explain) may appear more rational. Even for committed naturalists, the multiverse hypothesis may seem to be an overly speculative explanation of the data, and one that opens up the worrying possibility of appealing to a "chance of the gaps" (Dembski 2003) whenever we encounter data that does not fit into our hypothesis. After all, given a sufficiently large multiverse, there is bound to be a universe where just this data exists by chance, rather than because the data points to some factor that does not fit into one's worldview.

Biological Design

How does the situation look in biology? As noted, Paley's famous analogy of a watch on a heath argues that we would know that a watch we found on a heath is designed, and this is because of the watch's purposeful ordering, which distinguishes it from objects like stones. According to Paley, we would be able to know that the watch is designed, even if we had no prior knowledge of how watches are made. The reason is that "when we come to inspect the watch, we perceive (what we could not discover in the stone) that its several parts are framed and put together for a purpose." He then goes on to argue that "every indication of contrivance, every manifestation of design, which existed in the watch, exists in the works of nature; with the difference, on the side of nature, of being greater and more, and that in a degree which exceeds all computation" (Paley 2008 [1802], chapter 1). The order of the eye, the hand, the heart, and many other organs of human and animal bodies is among Paley's favorite examples of purposeful order in biology. It seems that the order of the eye, for example, is directed toward the purpose of enabling sight, and the order of the heart is directed toward the purpose of pumping blood. Together, all organs appear to have the purpose of sustaining the life of the animal.

In contrast with earlier design arguments, proponents of ID emphasize examples from the molecular level. Biochemist Michael Denton, whose work was a critical influence in the development of the early ID movement, argues that "nearly every feature of our own advanced machines [has] its analogue in the cell [...] so persuasive [is] the analogy, that much of the terminology we would use to describe this fascinating molecular reality [is] borrowed from the world of late twentieth-century technology" (Denton 1986, 329). According to Denton, the advance of molecular biology greatly strengthens the case for design: "It is the sheer universality of perfection, the fact that everywhere we look, to whatever depth we look, we find an elegance and ingenuity of an absolutely transcending quality, which so mitigates against the idea of chance" (Denton 1986, 342). The literature produced by proponents of biological design arguments is accordingly full of awe in the face of biology's ordered complexity, though it does also usually attempt to engage with some examples of the apparent imperfection of life as well (Kojonen 2016, 156–161).

While terms like purpose and teleology have often been treated with suspicion in biology, at the same time the biological literature is replete with references to functions, adaptations, and "teleolonomy," as well as comparisons to machinery and information science. Even though organisms are natural entities, it has proven difficult to understand these self-organizing, self-nourishing entities without teleological (directional, purposeful) language. The function of the heart seems to be to pump blood, the function of the lungs seems to be to breathe air, and so on, all of the organs working together to keep the animal alive. Due to the practical necessity and aptness of such language, there is a massive literature attempting to justify the use of these types of descriptions philosophically (Perlman 2004; Walsh 2008). While a full-blown teleological realism may be difficult to square with a naturalistic picture of the world (Leidenhag 2019), the appearance of teleology, at least, is thus difficult to deny.

For the purposes of this study, whatever the terminology, Darwinians do not deny the appearance of teleology in biology, but consider it one of the most important features of biology that is in need of explanation, as Ruse (2003) notes. In this the Darwinians continue in the tradition of Paley, to such an extent that even the most minute details of biology are often argued to have some function in aiding survival. Gould and Lewontin (1979) famously criticized this tendency as the "Panglossian paradigm," arguing that some features of life may be just evolutionary accidents. Recent ideas on neutral evolution and the idea of "the laws of biological form" from evolutionary developmental biology have continued this trend of minimizing the importance of function as the sole determining criterion of biological form. Nevertheless, none of these thinkers denies the importance of explaining the appearance of teleology.

That said, Darwinians can disagree with some proponents of design arguments about how to best describe these features. As noted, ID proponents commonly argue that life is based on biochemical "machinery." Behe (2006, x), for example, argues that "it was once expected that the basis of life would be exceedingly simple. That expectation has been smashed. Vision, motion, and other biological functions have proven to be no less sophisticated than television cameras and automobiles." However, some critics have argued that while machine metaphors are also commonly used in scientific journals, the metaphor bypasses central differences between life and machines, most crucially life's ability to evolve (Pigliucci and Boudry 2011; Nicholson 2019).

The situation is similar to the case of biological "information." ID proponent Stephen C. Meyer (2009) has prominently argued that the genetic information in DNA is a "signature in the cell," strong evidence of intelligent design in biology. Some concept of information is also generally thought to be essential for biology by mainstream scientists. Developmental biology is beginning to unravel the secrets of animal development, and how organisms use both genetic and epigenetic information to build the components they need to maintain themselves and reproduce (Noble 2008). However, there is still room to debate just what biological information is, and how it compares to the products of human programming. Philosopher of biology Peter Godfrey-Smith thus argues that ID's design argument is based on an overly robust idea of information as a non-material substance that cannot be reduced to physical processes, an idea which Godfrey-Smith (2007) argues goes beyond what is required by science. However, it seems to me that proponents of ID see the non-reducibility of information to physical processes as the conclusion of their argument rather than a premise. Similarly, the analogy with machines highlights the apparent purposefulness, sophistication, and amazing organization of life's complexity, which all agree is in need of explanation. In the ID movement's discourse, the word "machine" simply acts as the proper name for this type of order (Kojonen 2016, 51–54). But the appearance of design is enough to get the design argument started, into asking what explains the existence of this type of order.

St Thomas vs. Paley and ID

Whereas the aforementioned design arguments appeal to particular examples of seemingly purposeful order in nature, another possibility would be to argue instead based on the general existence of any teleology. Within the Christian tradition, the best-known version of the teleological argument dates to Aquinas, who presented it as one of his five main lines of argument for the existence of God:

> We see that things which lack intelligence, such as natural bodies, act for an end, and this is evident from their acting always, or nearly always, in the same way, so as to obtain the best result. Hence it is plain that not fortuitously, but designedly, do they achieve their end. Now whatever lacks intelligence cannot move towards an end, unless it be directed by some being endowed with knowledge and intelligence; as the arrow is shot to its mark by the archer. Therefore some intelligent being exists by whom all natural things are directed to their end; and this being we call God. (Aquinas, *Summa Theologiae*, I 2.3)

Here, Aquinas first argues that all natural things exhibit teleology, goal-directedness. For Aquinas, this teleology is not present in just biology, such as in the often used example of the acorn, which has the potential to grow into an oak tree in the right conditions. Rather, Aquinas follows Aristotle in arguing that teleology is present in everything: all things have their own tendencies to act in certain ways. In addition to living objects like the acorn, this includes nonliving objects like the moon, which has the tendency, *the telos*, to orbit the Earth. Teleology describes the potential of things to act in certain ways and to cause certain effects. But natural things cannot themselves be the sources of their own potentiality and directionality. Therefore, Aquinas concludes, "some intelligent being exists by whom all natural things are directed to their end."

In the contemporary discussion, many Thomists have argued that Aquinas' teleological argument differs in crucial respects from later design arguments (Newton 2014; Feser 2010). Aquinas' argument is based on the teleology of all things, not just a subset of things, like biological organisms. As such, the argument is not threatened by evolutionary biology, since evolution cannot explain the general orderliness of all things, but depends on it. As Newton (574) argues: "if some chance mutation proves to be beneficial it is because it occurs within an environment that is stable enough for that change to be selected: but this stability is nothing more than things having final causes."[4] The point is not that nothing beneficial can happen by chance. It can: for example, two people can fortuitously meet without planning to do so, or a farmer digging in his field can accidentally find a treasure that someone else buried there. Rather, as Feser (2009, 113) argues, Aquinas' argument modestly assumes only that "it is impossible that *every* apparent causal regularity can be attributed to chance, for chance itself presupposes causal regularity."

Accordingly, one way to defend the teleological argument in the contemporary discussion would be to advocate a return to robust Aristotelian-Thomistic metaphysics. However, the road back to such a metaphysics is difficult for many of us moderns. Perhaps it can indeed be argued that teleology is equally present in simple phenomena "like the attraction between two particles," which according to Feser (2009, 113, 2019) would suffice for Aquinas' argument, as in the order of biological organisms. However, insofar as we want to save the reliability of ordinary religious experience of nature, it does seem that some parts of nature are more evocative of design intuitions than others. One way to get back to a more non-reductive understanding of nature may indeed be to just affirm the general teleology of all of nature, but perhaps this journey can be aided by first recognizing the existence of particular examples of teleological order in nature, such as in biology.

Some other Thomists have argued for the compatibility of Aquinas' and Paley's approaches. George (2013, 2016) points out that Aquinas' argument in *Summa Contra Gentiles* seems to come closer to what Paley and the ID movement argue, with the emphasis on an extrinsic source for the order of separate parts of nature:

> It is impossible that things which are contrary and dissonant concord in one order always or for the most part, except by the government of someone, from which is assigned to each and every thing that it tend to a certain end; but we see in the world things of diverse natures agree in one order, not rarely and by chance, but always or for the most part. It is therefore necessary that there be something the providence of which governs the world; and this we call God. (Aquinas, *Summa Contra Gentiles* I, 13)

Paley's "contrivance" of biological organisms similarly means the ordering of multiple parts toward a common goal, and ID proponent Michael Behe (2006, 193) argues that design is primarily about the arrangement of parts for a purpose, and thus design serves well to explain the ordered complexity of biological cells. ID proponents see this ordering as ultimately deriving from outside the organism and its parts, since these parts are not intelligent. However, in Feser's (2010) understanding, Aquinas' argument includes a difference between the intrinsic teleology of natural objects and the extrinsic teleology of artifacts. For example, the parts of a plant have the inherent tendency to function together, whereas the parts of the artifacts must be arranged by their designers. But seeing plants as artifacts leads to seeing them as an accidental compilation of parts, leading to a problematic mechanistic understanding of nature.

In contrast to Feser, George (2016) claims that the difference between intrinsic and extrinsic teleology is not important for Aquinas' argument. Aquinas himself argues that natural objects tend toward their goals "as the arrow is shot to its mark by the archer," thus apparently equating the teleology of natural objects and man-made arrows. Nor do Paley and the ID movement necessarily emphasize just the extrinsic origin of teleology, or construe organisms only as machines, rather than as living entities (Turner 2017). Rather, the Paleyan design argument, and the ID versions, is plausible for the existence of a platonic extrinsic teleology on the basis of Aristotelian immanent teleology. As in the traditional mainstream Christian understanding, the teleology of organisms is intrinsic, but ultimately derives from a source outside these organisms, because of the doctrine of creation (Ariew 2002; Oliver 2016). All this helps bridge the gap between Thomistic and modern design arguments. Although the Paleyan tradition may have contributed to interpreting organisms as mere machines, without agency of their own, this does not seem like a necessary consequence of the design argument as such. Supposing that some aspect of biological order is well explained in terms of design does not mean that this aspect is the only thing that matters about biology.

Another proposed difference is that the Thomistic argument leads directly to God, whereas contemporary design arguments might also be compatible with a more limited creator. Since the purpose of the Thomistic arguments is to explain the teleological potential that is present in all things, the explanation cannot have such a potential in itself, but must be pure act, the "I AM," namely God (Newton 2014, 576). In contrast, contemporary versions of the design argument do not assume Thomistic metaphysics, and as such can provide only probabilistic supporting evidence for the existence of a Creator. However, I have already argued that such evidential support for religious belief is valuable, even without attaining to the status of proof.

3.2 The Logic of Design Arguments

Analogical, Deductive, or an Inference to the Best (or Only) Explanation?

How, then, does the design argument's logic work? Many assume that the biological design argument is an analogical argument, based on comparing machines and organisms. But I will argue that it is more plausible to understand it as a deductive inference (working from general principles to examine the specific case of biological teleology) or as an inference to the best explanation (comparing different possible explanations for biological teleology).[5]

I have already referenced Paley's watchmaker analogy, which Paley (2008 [1802], chapter 1) ends by arguing, "Every indication of contrivance, every manifestation of design, which existed in the watch, exists in the works of nature; with the difference, on the side of nature, of being greater and more, and that in a degree which exceeds all computation." As an analogical argument, this would be formulated approximately as follows:

> Premise 1. The ordered complexity of biology is similar to machines.
> Premise 2. Machines are designed.
> Premise 3. Similar effects typically have similar causes.
> Therefore, biology's ordered complexity is probably also designed.

Hume (2001 [1779] famously criticizes this argument. In his *Dialogues*, Hume focuses on the unreliability of premise 3. He acknowledges that analogies often lead to the correct conclusion. However, similarity is no guarantee of the conclusion, and analogies become less reliable when the compared objects are very different. For example, comparing humans and plants, we would be wrong to infer that since humans have blood, plants also have blood. There are great differences between the cosmos and machines, and between biological organisms and machines, and according to Hume, this makes the analogy unreliable. As Ratzsch and Koperski (2019) note, the criticism here is based on the realization that "any two (groups of) things have infinitely many properties in common and also differ in infinitely many respects." So why would sharing some properties necessarily imply that some other property is also shared, rather than one of the differences?

This represents analogical arguments as a comparison of the number of similarities in two cases, but the viability of analogies actually hinges on whether the similarities are relevant (Juthe 2005; McGrew 2004). Analogical arguments depend on some understanding of the causal dynamics at play. For example, though we would be wrong to conclude that plants have blood because they are somewhat similar to humans, we would be right to conclude that plants must also have some system to disperse nutrients in their bodies, since as living beings plants must also require sustenance. Proponents of design arguments, like Paley, similarly think that there is a causal connection between teleology and creative intelligence. Paley thus writes that "arrangement, disposition of parts, subservience of means to end, relation of instruments to a use, imply the presence of intelligence and mind" (Paley 2008 [1802], 9). In defense of Hume, the *Dialogues* also contains a similar response. The character Cleanthes argues that the cosmos resembles machines in the crucial respect of "adjustment of means to ends" (Hume 2001 [1779], II).

However, if the design argument is based on an understanding of the nature of teleology, and an understanding of the causal powers of minds, then it seems that the argument is not ultimately based on the comparison. Accordingly, some commentators have argued that Paley's design argument is actually a deductive argument or an inference to the best explanation, which the analogy merely illustrates (George 2016; Oppy 2002, but compare Schupbach 2005). As a deductive argument, we can formulate it as follows:

> Premise 1. Some things in nature (or nature itself) exhibit property D (such as complexity ordered toward a purpose)
> Premise 2. Property D is not producible by non-purposeful processes, but only by an intelligent being.
> Therefore, some things in nature (or nature itself) must have been purposefully created.

Here premise 2 corresponds to Aquinas' claim that "whatever lacks intelligence cannot move towards an end, unless it be directed by some being endowed with knowledge and intelligence," and Paley's claim that "arrangement, disposition of parts, subservience of means to end, relation of instruments to a use, imply the presence of intelligence and mind." Here, insofar as the premises are certain (or probable), the certainty or probability of the conclusion follows. However, to take account of the uncertainty, and the necessity of comparing explanations in such cases, it may be preferable to construct the design argument instead as an inference to the best explanation (Ratzsch and Koperski 2019), as it is also popular to do with respect to many other natural theological lines of argument (Banner 1990; McGrath 2016).

The basic logic behind the inference to the best explanation (Lipton 2003) is generally thought to be "abductive" as described by C. S. Peirce (1955, 151). The idea is that the ability of a hypothesis to explain some facts provides evidence in favor of the hypothesis—even if the hypothesis cannot be strictly speaking proven based on these facts. The basic logical structure of abductive inferences is as follows:

> Premise 1. The surprising fact C is observed.
> Premise 2. But if A were true, C would follow as a matter of course.
> Therefore, there is reason to suspect that A is true.

Many have argued that this type of logic is typical in scientific theory formation, and is also crucial for Darwinian evolutionary theory (Lipton 2003). Similarities constitute evidence in favor of the existence of a common ancestor, because this would explain the existence of these similarities as features inherited from this ancestor. Such arguments do not absolutely prove their conclusion—it could be that some other cause also explains the same facts. Thus we need some way to evaluate different explanations, and to judge which one is the best. Typically, adjudicating between competing explanations will require detailed work to produce evidence that one of these hypotheses explains better. For example, in the case of the similarities of living beings, rival explanations for individual similarities include common ancestry, convergent evolution (the independent evolution of the same feature more than once), and common design driven by the same functional requirements and the same design philosophy. The last of these is the explanation most commonly favored by proponents of Intelligent Design, although they can also accept common descent (Kojonen 2016, 56–60). But mainstream biologists also explain many instances of similarity by reference to convergence (McGhee 2011). It is the similarities that appear to be unnecessary for functional reasons, and the general pattern of similarity forming a "nested hierarchy" that, in the opinion of mainstream biologists, constitutes the crucial evidence making common descent a well-justified explanation. Additional criteria that come into play in evaluating competing explanations also include factors like the simplicity of the hypothesis (which contributes to its initial plausibility), its fruitfulness in also helping to explore and answer other questions, and its fit with other ideas that we already accept. There is a long list of such criteria (Lipton 2003). If one explanation is superior based on the criteria that we have adopted, then it is seen as justified to adopt this as the probably true one—at least until a better one emerges.

Lipton (2003, 56–66) has distinguished potential and actual explanations. All explanations that explain the data in some way can be called potential explanations, but the actual explanation also needs have the additional property of being true. In the case of apparently designed objects in nature, potential explanations have included chance as coincidence (as in a cloud's shape accidentally resembling a weasel), evolution, and design. Insofar as design is at least a potential explanation, then the evaluation could at least in principle result in design also being the best explanation. In recent discussion, the inference to the best explanation has been a popular way of formulating various design arguments (Collins 2009; Meyer 2009). One benefit of the approach is that as an inference to the best explanation, the design argument would not require the absolute elimination of alternative explanations.[6]

The design argument as an inference to the best explanation (IBE) is typically formulated roughly as follows:

> Premise 1. Some things in nature (or nature itself) exhibit property D (such as complexity ordered toward a purpose)
> Premise 2. This property would be well explained if design was the cause. However, other explanations explain this property poorly.
> Therefore, when comparing all available explanations, design is the best overall explanation of this property.
> Therefore, property D was probably designed; at least this property provides more reason to infer design over competing explanations.

My above formulation of the design argument is an IBE to state that "at least this property provides evidence that supports design over competing explanations." However, in practice, proponents of probabilistic design arguments often argue that the conclusion is so certain that it comes at least close to a proof of design. This brings their inference to the best explanation closer to what Bird (2005, 2007, 2010) has called the "inference to the only explanation." Bird terms this "Holmesian inference," following the logic of Sherlock Holmes: "when you have eliminated the impossible, whatever remains, no matter how improbable, must be the truth" (Doyle 2001[1892]). However, as Bird (2005, 26–28) emphasizes, this is a demanding form of inference whose criteria are not fulfilled even by all rational inferences in the natural sciences or in everyday life. The problem of underdetermination often prevents this, since an explanation cannot be the only one if not all possible explanations can be decisively eliminated. Typically, it will be sufficient to show that rival explanations are improbable, without the need for decisively eliminating them.[7]

Because the deductive formulation can also be understood probabilistically (so that even less than fully certain premises will give some support for the conclusion), and the inference to the best explanation can be developed toward being an inference to the only explanation, both versions of the argument end up being rather similar. In both arguments, premise 2 is the central controversial premise. It can be attacked by either denying the explanatory power of design (Does design really explain?), or by presenting different explanations for the supposedly designed properties (Do we have a better explanation?). Additionally, features like the background probability of design (how well design fits or does not fit with our other background beliefs) are important. To take these into account more formally and to evaluate their precise effect on the credibility of the design argument, it could be useful to formulate the design argument as a Bayesian inference (McGrew 2004; Barnes 2018, Swinburne 2004a, appendix). But in this chapter it will suffice to simply note the influence of these factors, and then move on to assess the effect of evolutionary biology on the argument. I now briefly assess philosophical critiques through thought experiments.

Thought Experiments and Objections

In the previous discussion, I have already noted three philosophical objections to the credibility of design as an explanation. First, it is argued that we can only reliably detect human design because our design detection is based on our experience of human actions and artifacts, which cannot be generalized to a supposed non-human, supernatural, or divine designer. Second, as a supporting reason for being able to detect only human design, it is argued that design detection requires knowing about the existence and motives of the designer beforehand, as well as knowledge of viable mechanisms that the designer could have used. Since (it is alleged) the capabilities and motives of God are beyond human knowledge, we cannot formulate a design argument. Third, it is argued that design fails as an explanation, because it opens the additional question of "who designed the designer."

Sober (2000, 32–33) illustrates the difficulties facing design arguments with the following thought experiment. Suppose that we hear a strange rumbling noise from the attic. We could argue that the hypothesis "there are gremlins in the attic, and they are bowling" explains the noise quite well, since it predicts the observed empirical evidence. This is a causally sufficient explanation: bowling is indeed experientially known to cause a lot of noise, and we can well easily formulate and understand the concept of a gremlin bowler. However, as Sober notes, the explanation still does not feel satisfactory. Two factors help understand why. First (1), we do not have any independent reasons for believing in gremlins, which means that the prior probability of the gremlin hypothesis is extremely low. Thus, though the evidence increases the probability of the gremlin hypothesis, it is not sufficient to raise the probability into the realm of credibility. Second (2), almost any data can be similarly "explained" by appealing to gremlins with just the right kind of motivation and powers. For example, supposing that cookies are missing from the cookie jar, we could hypothesize that an invisible gremlin ate them. But all such explanations seem too ad hoc, made up to explain just this one piece of data, and are without independent support.[8]

According to Sober, design as an explanation for life's order also has a very low prior probability and is similarly ad hoc. In order for design to be credible, we would need some plausible reason for supposing that the designer exists, and has the capabilities and the motivation for producing the order we see. Thus, while Sober rejects many of Hume's critiques of the design argument (because they apply only to the analogical form of the argument), his conclusion is still very Humean. According to Sober, our inferences about divine design are not reliable, since we only have experience of human design. As he summarizes elsewhere: "The more our hypotheses of intelligent designers depart from the human case, the more in the dark we are as to what the ground rules are for inferring intelligent design" (Sober 2003, 38). Similarly, Wilkins and Elsberry (2001) and Narveson (2003) argue that we cannot explain by reference to cosmic designers, because their actions are beyond our ability to predict: "Bodiless minded super-creators are a category that is way, way out of control" (Narveson (2003, 99). Indeed, it seems to be an empirically established fact that background beliefs have a great effect on design beliefs and the lack of such beliefs. Even the possibility of perceiving design-like patterns requires such knowledge (De Cruz and De Smedt 2010).

Here proponents of design arguments have two main ways of responding. The first way is to assert that we can have moral knowledge, and this makes it possible to narrow down what a good Creator would be expected to do, even though we cannot expect to predict any personal agent's actions precisely, nor should we demand such ability for exact predictions from personal explanations. In addition, within a Christian theology of nature, we can have reason to think that God has certain motives, and the design argument can then provide evidence that is consonant with the theology of creation. In such a framework, there is no reason to think that our understanding of God must be conjured out of thin air. Rather, we are evaluating whether the evidence is consonant with a specific theology that we have independent reason to believe (Abraham 1985, 84).

The second way is to argue that there is some property which is strongly linked to intelligent agency, and that this can help make the design hypothesis far more plausible than the gremlin hypothesis. As I have argued elsewhere (Kojonen 2016, 139), the gremlin thought experiment itself can be modified to demonstrate this. Suppose the noise from the attic was like complex orchestra music, rather than just apparently random rumbling. In that case, design would be the best explanation, even if we had no prior evidence of the designer, simply based on the pattern that we observed. Because of our background beliefs, the most reasonable possibility would be a human designer. However, if this were decisively ruled out (e.g., if the music predated human existence), then a non-human designer would be the reasonable conclusion, because the link between music and minds is so strong. Hence, Sober's thought experiment does show that the background probability of the existence of a designer does affect our conclusions about design. However, it seems to be stretching the point too far to argue that prior knowledge of the existence of a designer is absolutely necessary. If we understand the search for the best explanation as comparing competing hypotheses, then it is clear that the evidence in this case would provide powerful support for the hypothesis of design over all other possible explanations, even if there was no way of predicting the existence of such music beforehand (or deriving it as a certain consequence from the hypothesis of a designer).

Thought experiments, like Sober's example of noise in the attic, are one of the traditional ways of bringing out the implications of different philosophical ideas. They are the "laboratories of the mind" for deciding between competing ideas, and defenders of design arguments commonly appeal to them to justify the possibility of detecting non-human design (Brown and Fehige 2019).[9] There is nothing strange about this philosophically. Thought experiments are usually thought to be essential to philosophy, but they have even played a surprising role in natural science, from Galileo to quantum physics (Palmerino 2011; Hossenfelder 2013). Whereas scientific experiments provide a way to test our theories against the reality of nature, thought experiments test the logical coherence and intuitive plausibility of our ideas.

Despite the popularity of thought experiments in philosophy, it could be argued that intuitions are generally unreliable, and thus thought experiments that test our intuitions do not yet tell us much (Peels et al., ed. 2020). However, it seems to be the case that intuitions about explanations, at least, continue to be important in the philosophy of science and scientific practice. As Ylikoski (2001, 51) argues, "science has its origin in common sense cognition. There should be some kind of continuity. Of course, it now includes much that is apparently incompatible with common sense, but I seriously doubt that this incompatibility also extends to the general principles of explanatoriness." There is no way to test philosophical principles of explanation scientifically in any way that would not itself rely on some philosophical principles of explanation. Thus, it seems that philosophical intuitions about explanatoriness, at least, should be taken seriously, if not necessarily uncritically. Accordingly, thought experiments regarding the possibly explanatory power of design also deserve to be noted.[10]

In my view many thought experiments presented by proponents of design arguments do help show that detecting non-human design is possible in principle (Kojonen 2016, chapter 7). I will now consider a few examples. First, recall Paley's (2008 [1802], chapter 1) example: "suppose I had found a watch upon the ground, and it should be inquired how the watch happened to be in that place." This part of Paley's example is usually quoted, but his continuation of the argument is also important. Step by step, he begins to modify the example, asking us to suppose, for instance, that we do not have prior knowledge of watches, and that the watches were capable of reproducing. Through these modifications, he uses the example to undercut possible responses to the design argument, chiefly related to the differences between life and watches. Paley also argues against several more philosophical objections: he argues, for example, that in the case of the watch, it would be absurd to assert that we should simply accept the existence of its order as a brute fact that is incapable of being explained further. His implied conclusion is that it would be similarly absurd to assert that life is just a brute fact, and the natural stopping point in our search for explanations. Order oriented toward a purpose cries out for an explanation in the case of the watch, and can be explained by reference to a mind, so why could such an explanation also be legitimate in the case of nature?

Contemporary defenders of design arguments similarly use such analogies or thought experiments to respond to philosophical critiques against the argument. For example, Leslie's influential defense of cosmic design argument in his book *Universes* (1985) utilizes thought experiments in practically every chapter. One problem with many of the commonly used critiques is that they seem to prove too much: if the critique was correct, and the logic of the design argument was faulty, then how can we explain the intuitively obvious true detection of design in these examples?

Leslie notes the objections that fine-tuning cannot provide evidence of design, because (1) this is the only universe we have observed, (2) it is unreasonable to seek an explanation for the properties of the laws of nature, since the laws are themselves the fundamental reality, and (3) there is nothing mysterious about the fact that natural laws allow for life because otherwise we could not observe them. However, according to Leslie, the question-begging nature of these objections becomes clear if we consider a hypothetical scenario in which "particles regularly formed long chains which spelled out 'GOD CREATED THE UNIVERSE', this then being shown to result inevitably from basic physics" (Leslie 1989, 109). It seems clear that it would hardly be compelling to claim that such a pattern is not evidence of design, because this is the only universe that we have observed. Nor would it be reasonable to claim that this order is just a brute fact that we should not even attempt to explain, or to claim that our observation of the pattern suffices as the explanation. Leslie's point is that if the objections against the design argument work against fine-tuning, then they should also work in the situation described by the thought experiments, all else being equal. However, it is intuitively absurd that such a pattern could not give any evidence of design. Therefore, there must be something wrong in the aforementioned objections to the fine-tuning design argument.

Within the ID movement, appeals to examples from science fiction are common (Kojonen 2016, 113–115). The most commonly referenced example is the *Search for Extraterrestrial Intelligence* project (SETI).

Proponents of ID argue that if an information-rich message were received from outer space, such as the one described in SETI pioneer Carl Sagan's science fiction book *Contact* (1985), then this would provide clear evidence of design. Accordingly, they argue, prior knowledge of a designer's existence is not required before making a design inference. These ID proponents then go on to argue that just the same type of order is also found in nature, and this order thus provides evidence that life is intelligently designed (e.g., Dembski 1998a).

Some responses to such thought experiments seem to miss the point. Pennock, for example, argues against Dembski's use of such examples by stating that "in each case we infer an intelligent signaller not because these are cases of complex specified information in a generic sense, but because the pattern of information matches a previously known pattern that we associate with intelligence"(Pennock 1999, 254). SETI researcher Shostak (2005) has also criticized the use of SETI as an analogy to biological design arguments. Shostak argues that SETI actually does not investigate the content of signals at all (as in Sagan's *Contact*) but focuses on finding an intensive narrow-wavelength signal. These sorts of signals, Shostak argues, are known to be produced only by civilizations. However, in both cases it seems that the critics affirm the crucial point of the thought experiments: a pattern that we have reason to associate with intelligence can indeed provide evidence of non-human, previously unknown intelligence. These critics do also argue that biological teleology does not provide such evidence, since we can explain it via Darwinian processes. That is indeed relevant for ID's argument, but does not refute the point of the thought experiments. Moreover, if Darwinian explanations are actually compatible with the biological design argument, then these differences do not refute the argument at all.

In my view such thought experiments do succeed in showing that we could in principle have good evidence of non-human design in nature. However, it remains possible for the critic to argue that our situation is different from the hypothetical situations of the thought experiments, and that in our actual situation we do not have evidence of design—or at least not equally strong evidence of design. In nature, we do not find "GOD CREATED THE UNIVERSE" patterns on the subatomic level. We do find astoundingly complex teleological patterns, but the evidence for design is not at quite the same level as in the thought experiments. This means cultural ideas about nature and prior beliefs about the existence of

God will inevitably have a greater effect on the persuasiveness of the design argument.

Following Sober's example, a committed atheist who believes the prior probability of the design hypothesis to be extremely low will require more evidence for the conclusion of design than someone who believes at least in the possibility of the existence of God. For the religious believer and the natural theologian, the existence of God is not ad hoc at all—and theists will typically argue that atheists give an unjustifiably low prior probability to God's existence and design. However, just because the persuasiveness of the evidence is affected by our worldviews, this does not mean that the evidence does not exist, or that it does not support the conclusion of design at least to some degree. It would be strange to argue that evidence for design has to be as clear as in the cases of the thought experiments before we can have any evidence at all, or before the evidence can rationally support belief in design. Perhaps the amount of evidence and the quality of arguments we have is instead somewhere between the extremes of "overwhelming/undeniable" and "nonexistent." Again, suppose that the understanding of Pascal and Evans, described in Chap. 2, is correct, and there is theological reason to suppose that any evidence of God should be resistible. It would seem unfortunate if such an understanding were ruled out by principles of reasoning that would force us into the binary choice of either having undeniable evidence of design or no evidence at all.

The thought experiments, moreover, do not yet demonstrate what, precisely, is wrong with the objections. Rather, they only show that something is likely to be wrong, since the objection leads to an absurd and implausible conclusion.[11] For example, in the case of the old question "Who designed the designer" should this really stop design arguments? Should we think that, say, the "GOD CREATED THE UNIVERSE" message gives no evidence of design, because we can then still ask "Who created God?" This would seem strange, though it does not yet show what, precisely, is wrong with the objection. But perhaps it will open us to seeking answers from the history of theology, according to which the question is meaningless. There are many reasons why God as a necessary and self-sufficient being would be a more natural stopping point for the search for explanations than contingent beings (Feser 2017, 249–268; Glass 2012). Proponents of minimalistic design arguments, such as proponents of ID, cannot use such arguments without first committing themselves to a more robust doctrine of creation. They will instead have to argue that an explanation can be well supported by the evidence even if it raises additional questions.

Thus, thought experiments provide reason to think that there is something wrong with the common Humean objections, and that it is indeed in principle possible to have evidence of design. This emphasizes the importance of analyzing the evidence and determining the viability of design arguments based on the evidence, rather than ruling out design on a priori philosophical grounds. Indeed, the critics of the aforementioned thought experiments also emphasize the importance of natural explanations for biological teleology. This, for them, creates a core difference between the situation of the thought experiments and biology, and this is also what I will mostly focus on in the coming chapters. Hence, though the case here may still be incomplete, I will simply ask the reader to assume with me that philosophical objections to design arguments are not sufficient without added Darwinian objections. Supposing that this is the case, then how do natural explanations in general, and evolution in particular, affect the picture? This is what I will be asking in the chapters that follow.

3.3 Conclusions

Although the idea that nature is designed is intuitively plausible to many people, design arguments have also come under philosophical and scientific critique. In this chapter, I have argued that many of these critiques are based on principles of reasoning that would not allow us to discern evidence of design even in cases where this evidence was seemingly very compelling. Rather, adopting such principles of reasoning would mean commitment to a "naturalism of the gaps," where the truth of naturalism is assured already based on a priori considerations, before examining any evidence. While methodological naturalism may well be justified as a principle of scientific inquiry (because the sciences must limit the scope of their inquiry in order to be effective), naturalism cannot be assumed as a general principle of all rational discussion of worldviews. This would mean that we would be unable to ever recognize evidence of design, even supposing that the universe really is purposefully created. But what if the universe really is designed? In that case we could never recognize it. In practice, such a strong insistence on the sufficiency of philosophical critiques of natural theology seems to lead to a "naturalism of the gaps" that is every bit as problematic as the commonly criticized God of the gaps reasoning. At least, it seems far preferable to criticize design arguments also on a posteriori, rather than just a priori, grounds. And evolution is the most important ground for such a critique.

That said, the logic of design arguments deserves more analysis than has been possible here. The purpose of this brief analysis has been simply to argue that the question of design is serious, and it is at least not clear that the philosophical objections are sufficient to overturn the claim that biological teleology and cosmic fine-tuning provide evidence of design. It may be that different conclusions about design in the cosmos will ultimately depend on different answers to metaphysical and philosophical questions, such as the proper stopping point of the search for explanations. If one could argue oneself into believing in the existence of a cosmos, with its fine-tuned laws of nature, as a brute fact, eschewing the search for any explanation, then this might help resist the conclusion of design. It may well be that if the question of origins is admitted as legitimate, then resisting the conclusion of design based on the characteristics of the cosmos becomes difficult. Nevertheless, even if the design argument supports religious belief in God, it could be that there are other arguments, such as the argument from evil, which count against it.[12]

This chapter has also highlighted the context-dependency of the conclusion of design. Considered in isolation, the evidence of design-like patterns in nature does provide supporting evidence for belief in the existence of a Creator. Nevertheless, for someone for whom the idea of a Creator is as strange as the idea of "gremlins bowling in the attic," the evidential support given by the design argument may not be sufficient to compel belief in the Creator by itself. While there are different estimations of the strength of the evidence of design, we can at least imagine much stronger (and much weaker) evidence of it. Thus, while the evidence of design supports belief in purpose, I will grant that the conclusion of design is nevertheless easier to accept for those who also have other reasons to believe. But this in itself does not indicate that the evidence of design is weak; it may also be that the underlying assumption about the extreme implausibility of theism is at fault. Moreover, there is no reason to think that a design argument needs to compel belief in order to be a good argument.

Yet estimating the precise strength of the evidence of design is not the purpose of this book, but rather the assessment of the compatibility of design and evolution as explanations. It is not difficult to find critics of design arguments who would admit that if they had lived before Darwin, then they too would have found the case for design compelling, simply on the basis of the intricate orderliness of living organisms. In their view, it is evolution, not philosophical considerations, which is central for criticizing the design argument. And as noted, even some of the most skeptical philosophical critics of design arguments have admitted that nature's order provides us with intuitively compelling evidence of design, and that the appeal of the idea of design cannot thus be wholly dispelled by philosophical considerations. For these reasons, philosophers and theologians critical of design arguments also typically rely on evolution as a counterargument to design, in addition to philosophical critique. This all speaks to the power of the evidence of design.

Based on the analysis of the logic of the design argument in this chapter, the idea of evolution as a defeater does have some initial plausibility. Biological design arguments commonly rely on an inductive or conceptual connection between a certain type of order (such as complex teleology) and purposeful design as a cause. Traditionally, proponents of design arguments have claimed that this type of order is impossible, or extremely difficult to explain without reference to design. In this chapter, I have already shown how evolution is argued to undercut this premise of the argument by showing how evolutionary processes can produce teleology. In the case of the analogical formulation of the design argument, evolution might be seen as a relevant difference between biological organisms and human artifacts, making the analogy unreliable. In the case of deductive design arguments, evolution is seen to challenge the idea that only design can explain the appearance of teleology (or other features suggestive of design). In addition, the case of the inference to the best explanation, design, and evolution are commonly presented as competing explanations. From such an understanding, critiquing evolutionary explanations would raise the status of design, and the viability of evolution makes design unnecessary.

However, this shows that the opposition of design and evolution may also be partly an illusion resulting from how the inference is constructed. In the case of complex phenomena, it is often the case that there is not just a single "best explanation," but rather different facets of the phenomena are explained by different explanations. Obtaining the full explanation may require combining, rather than just contrasting, explanations. Moreover, proponents of design—from St Thomas to Paley to Asa Gray to contemporary advocates—do allow for the possibility that designers can work through secondary causes, while retaining explanatory power. Seemingly, this might provide grounds for the compatibility of natural causes and design also in the case of biology. In the field of science and religion, the idea of the complementarity of explanations has also received attention (McGrath 2019). What, then, is the nature of evolutionary explanations of teleology, and why should the existence of such explanations refute design? It is to these questions that I turn in the next chapter.

Notes

1. Swinburne's discussion largely still assumes the old picture of deductive-nomological explanations as the norm in science. But this critique does not undermine Swinburne's basic point that personal explanations can be explanatory even if they are different from scientific ones.
2. There is debate about whether intentional explanations should be considered as causal explanations. In the following, I treat them as causal explanations, but for the contrasting view, see Sehon (2005).
3. Collins argues that this holds for restricted multiverse hypotheses where the number of universes is not infinite and eternally existing, not for unrestricted multiverses where all logically possible worlds exist. But an infinite multiverse is highly problematic for other reasons, such as the Bolzmann brain problem.
4. Reiss (2009, 140) agrees that "Darwin introduced a teleological determinism into the heart of his theory. This teleology is expressed in two related conceptions: (1) that evolution is a process going from a less-adapted to a better-adapted state and (2) that natural selection is a deterministic force, or agent, that directs the evolutionary process toward this better-adapted state." Reiss' response is to argue for reinterpreting Darwin's theory to remove the teleological elements, which in his view do not belong to this scientific theory.
5. A more extensive analysis of the matter can be found in Kojonen (2016, chapters 6, 7 and 8).
6. Behe's (2006, 256) design argument is also probabilistic. However, he is open to several different logical formulations of the argument. I have argued for understanding ID's design arguments as utilizing IBE in Kojonen (2016, 135). The most importance exception to the trend is Dembski (1998b), who argues that design is detected by eliminating chance and law-like regularities as possible explanations by using the criterion of specified complexity. However, Dembski does also claim that we can in addition inductively link specified complexity and intelligence (Kojonen 2016, 130–133). It seems that such a positive link is essential for his argument to work, and it seems that something like the IBE is needed to supplement Dembski's argument. See further Loikkanen (2019).

7. Elsewhere (Kojonen 2016, 145–147) I have accordingly argued the design argument could be a good and valuable argument even if alternative explanations are not eliminated.
8. Kitcher (1981) called this the "problem of spurious unification." The problem is that if a pattern of explanation could fit any state of affairs, it cannot explain why this particular state obtains instead of any other. Similarly also Pennock (1999, 275).
9. Note that the thought experiments here are of a commonly accepted type, as they simply test our intuitions about explanations, and do not aim at proving the existence of metaphysical entities by linking conceivability and possibility.
10. Here I am using Pust's (2019) definition of rational intuitions as "mental states in which a proposition seems true."
11. To elaborate further, I would also argue that we can often be more certain *that* we know something than *how* we know this. Williamson (2000) famously defended a "knowledge-first" epistemology, arguing that epistemology should begin with clear examples of knowledge, and then construct an epistemological theory (insofar as this is even possible) that accounts for these examples. Following critics, I would want to say that it is important to understand epistemological principles (McGlynn 2014). Nevertheless, it seems to me that a theory of knowledge should at least be able to account for intuitively clear instances of knowledge, and similarly a theory of design detection should be able to make sense of the intuitively clear design detection in the case of the thought experiments. Many common critiques of design arguments presuppose a theory of design detection that is unable to make sense of these examples, and this is a problem for these critiques.
12. On refusing the question, and how the theist might respond, see Feser (2017, chapter 7) and Turner (2004, chapter 12).

References

Abraham, William. 1985. *An Introduction to the Philosophy of Religion*. Englewood Cliffs, NJ: Prentice-Hall.

Ariew, André. 2002. Platonic and Aristotelian Roots of Teleological Arguments. In *Functions: New Essays in the Philosophy of Psychology and Biology*, ed. André Ariew, Robert Cummins, and Mark Perlman, 7–32. New York, NY: Oxford University Press.

Banner, Michael C. 1990. *The Justification of Science and the Rationality of Religious Belief*. Oxford: Oxford University Press.

Barnes, Luke A. 2013. The Fine-Tuning of the Universe for Intelligent Life. *Publications of the Astronomical Society of Australia* 29 (4): 529–564.

———. 2018. Fine-Tuning in the Context of Bayesian Theory Testing. *European Journal for Philosophy of Science* 8 (2): 253–269.

Barrow, John D., and Frank Tipler. 1986. *The Anthropic Cosmological Principle*. Oxford: Clarendon Press.

Behe, Michael J. 2006. *Darwin's Black Box: The Biochemical Challenge to Evolution*. 10th Anniversary ed. New York, NY: The Free Press.

Bird, Alexander. 2005. Abductive Knowledge and Holmesian Inference. In *Oxford Studies in Epistemology*, ed. Tamar Szabo Gendler and John Hawthorne, 1–31. Oxford: Oxford University Press.

———. 2007. Inference to the Only Explanation. *Philosophy and Phenomenological Research* 74: 424–432.

———. 2010. Eliminative Abduction: Examples from Medicine. *Studies in History and Philosophy of Science* 41: 345–352.

Brown, James Robert, and Fehige, Yiftach. 2019. Thought Experiments. *The Stanford Encyclopedia of Philosophy*. Edward N. Zalta (ed.). Available at https://plato.stanford.edu/archives/win2019/entries/thought-experiment/.

Cleland, Carol. 2011. Prediction and Explanation in Historical Natural Science. *British Journal of the Philosophy of Science* 62 (3): 551–582.

Collins, Robin. 2006. A Critical Evaluation of the Intelligent Design Program: An Analysis and a Proposal. Unpublished Paper. Available at http://home.messiah.edu/~rcollins/Intelligent%20Design/INTELL3.htm.

———. 2009. The Teleological Argument. In *The Blackwell Companion to Natural Theology*, ed. William Lane Craig and J.P. Moreland, 201–281. Malden, MA: Wiley-Blackwell.

———. 2018. The Argument from Physical Constants: Fine-Tuning for Discoverability. *Two Dozen (Or So) Arguments for God's Existence*, ed. Jerry L. Walls & Trent Dougherty, 89–107. Oxford: Oxford University Press. A previous draft is available at https://pdfs.semanticscholar.org/6ab2/7366b32edcf3d12587a07397e0d1d8e496fd.pdf.

Dawes, Gregory. 2009. *Theism and Explanation*. New York, NY: Routledge.

Dawkins, Richard. 1986. *The Blind Watchmaker*. Harlow: Longman.

De Cruz, Helen, and Johan De Smedt. 2010. Paley's IPod: The Cognitive Basis of the Design Argument Within Natural Theology. *Zygon* 45 (3): 665–684.

———. 2015. *A Natural History of Natural Theology: The Cognitive Science of Theology and Philosophy of Religion*. Cambridge, MA: The MIT Press.

Dembski, William A. 1998a. Introduction: Mere Creation. In *Mere Creation: Science, Faith and Intelligent Design*, ed. William A. Dembski, 13–32. Downers Grove, IL: InterVarsity Press.

———. 1998b. *The Design Inference: Eliminating Chance through Small Probabilities*. Cambridge Studies in Probability, Induction and Decision Theory. Cambridge: Cambridge University Press.

———. 2003. The Chance of the Gaps. In *God and Design: The Teleological Argument and Modern Science*, ed. Neil A. Manson, 251–274. London: Routledge.
Denton, Michael. 1986. *Evolution: A Theory in Crisis*. Bethesda, MD: Adler & Adler.
Doyle, Arthur Conan. 2001 [1892]. The Adventure of the Beryl Coronet. In *The Adventures of Sherlock Holmes*. London: Electric Book Co.
Feser, Edward. 2009. *Aquinas: A Beginner's Guide*. Oxford: Oneworld.
———. 2010. Teleology: A Shopper's Guide. *Philosophia Christi* 12: 142–159.
———. 2017. *Five Proofs of the Existence of God*. San Francisco, CA: Ignatius Press.
———. 2019. *Aristotle's Revenge: The Metaphysical Foundations of Physical and Biological Science*. Neunkirchen-Seelscheid: Editiones-Scholasticae.
George, Marie. 2013. What Would Thomas Aquinas Say about Intelligent Design? *New Blackfriars* 94 (1054): 676–700.
———. 2016. Thomistic Rebuttal of Some Common Objections to Paley's Argument from Design. *New Blackfriars* 97 (1069): 266–288.
Geraint, Lewis F., and Luke A. Barnes. 2016. *A Fortunate Universe: Life in a Finely Tuned Cosmos*. Cambridge: Cambridge University Press.
Glass, David. 2012. Darwin, Design and Dawkins' Dilemma. *Sophia* 51 (1): 31–57.
Godfrey-Smith, Peter. 2007. Information in Biology. In *The Cambridge Companion to the Philosophy of Biology*, ed. David L. Hull and Michael Ruse, 103–119. Cambridge: Cambridge University Press.
Gonzales, Guillermo, and Jay W. Richards. 2004. *The Privileged Planet: How Our Place in the Cosmos Is Designed for Discovery*. Washington, DC: Regnery Publishing.
Gould, Stephen Jay, and Richard Lewontin. 1979. The Spandrels of San Marco and the Panglossian Paradigm: A Critique of the Adaptationist Programme. *Proceedings of the Royal Society. Series B* 205 (1161): 581–598.
Halvorson, Hans. 2018. A Theological Critique of the Fine Tuning Argument. In *Knowledge, Belief, and God: New Insights in Religious Epistemology*, ed. Matthew A. Benton, John Hawthorne, and Dani Rabinowitz, 122–135. Oxford: Oxford University Press.
Hossenfelder, Sabine. 2013. Minimal Length Scale Scenarios for Quantum Gravity. *Living Reviews Relativity*. 16 (2): 11–22.
Hume, David. 2001 [1779]. *Dialogues Concerning Natural Religion*. South Bend, IN: Infomotions, Inc.
Juthe, A. 2005. Argument by Analogy. *Argumentation* 19: 1–27.
Kitcher, Philip. 1981. Explanatory Unification. *Philosophy of Science* 48: 507–531.
Kojonen, Erkki Vesa Rope. 2016. *The Intelligent Design Debate and the Temptation of Scientism*. London: Routledge.
———. 2017. Methodological Naturalism and the Truth Seeking Objection. *International Journal for Philosophy of Religion* 81 (3): 335–355.

Koons, Robert C. 2017, April 22. Are Probabilities Indispensable to the Design Inference? Unpublished Paper. Available at http://robkoons.net/media/3d211414d9a8a675ffff80c3ffaf2815.pdf.

Leidenhag, Mikael. 2019. Does Naturalism Make Room for Teleology? The Case of Donald Crosby and Thomas Nagel. *American Journal of Theology & Philosophy* 40 (1): 5–19.

Leslie, John. 1989. *Universes*. London: Routledge.

Lipton, Peter. 2003. *Inference to the Best Explanation*. London: Routledge.

Loesberg, Jonathan. 2007. Kant, Hume, Darwin, and Design: Why Intelligent Design Wasn't Science Before Darwin and Still Isn't. *The Philosophical Forum* 38 (2): 95–123.

Loikkanen, Juuso. 2019. *A Bridge between Science and Theology? William A. Dembski's Theory of Intelligent Design*. Publications of the University of Eastern Finland. Dissertations in Education, Humanities, and Theology. No. 140.

Loke, Andrew. 2017. *God and Ultimate Origins: A Novel Cosmological Argument*. Basingstoke: Palgrave.

McGhee, George. 2011. *Convergent Evolution: Limited Forms Most Beautiful*. Cambridge, MA: The MIT Press.

McGlynn, Aidan. 2014. *Knowledge First?* Basingstoke: Palgrave.

McGrath, Alister. 2016. *Re-Imagining Nature: The Promise of a Christian Natural Theology*. Oxford: Wiley-Blackwell.

———. 2019. *The Territories of Human Reason: Science and Theology in an Age of Multiple Rationalities*. Oxford: Oxford University Press.

McGrew, Lydia. 2004. Testability, Likelihoods and Design. *Philo* 7 (1): 5–21.

McGrew, Timothy, Lydia McGrew, and Eric Vestrup. 2003. Probabilities and the Fine-Tuning Argument: A Skeptical View. In *God and Design. The Teleological Argument and Modern Science*, ed. Neil A. Manson, 200–208. London: Routledge.

Meyer, Stephen C. 2009. *Signature in the Cell: DNA and the Evidence for Intelligent Design*. San Francisco, CA: HarperOne.

Narveson, Jan. 2003. God by Design? In *God and Design: The Teleological Argument and Modern Science*, ed. Neil A. Manson, 88–103. London: Routledge.

Newton, William. 2014. A Case of Mistaken Identity: Aquinas's Fifth Way and Arguments of Intelligent Design. *New Blackfriars* 95 (1059): 569–578.

Nicholson, Daniel J. 2019. Is the Cell Really a Machine? *Journal of Theoretical Biology* 477: 108–126.

Noble, Denis. 2008. *The Music of Life: Biology Beyond the Genome*. Oxford: Oxford University Press.

Oliver, Simon. 2016. Augustine on Creation, Providence and Motion. *International Journal of Systematic Theology* 18 (4): 379–398.

Oppy, Graham. 2002. Paley's Argument for Design. *Philo* 5: 161–173.

Orr, Allen. 2007. A Mission to Convert. Review of '*The God Delusion*'. *The New York Review of Books*, January 11.

Paley, William. 2008 [1802]. *Natural Theology*. Oxford World's Classics. Oxford: Oxford University.

Palmerino, Carla Rita. 2011. Galileo's Use of Medieval Thought Experiments. In *Thought Experiments in Methodological and Historical Contexts*, ed. Katerina Ierodiakonou and Sophie Roux, 101–126. Leiden: Brill.

Peels, Rik, Jeroen de Ridder, and René van Woudenberg, eds. 2020. *Scientific Challenges to Common Sense Philosophy*. London: Routledge.

Peirce, C.S. 1955. *Philosophical Writings of Peirce*. Ed. Justus Buchler. New York, NY: Dover.

Pennock, Robert. 1999. *The Tower of Babel: The Evidence against the New Creationism*. Cambridge, MA: MIT Press.

Perlman, Mark. 2004. The Modern Philosophical Resurrection of Teleology. *The Monist* 87 (1): 3–51.

Philipse, Herman. 2012. *God in the Age of Science? A Critique of Religious Reason*. Oxford: Oxford University Press.

Pigliucci, Massimo, and Maarten Boudry. 2011. Why Machine-Information Metaphors Are Bad for Science and Science Education. *Science & Education* 20 (5–6): 453–471.

Plantinga, Alvin. 2007. The Dawkins Confusion: Naturalism ad Absurdum. *Books & Culture. A Service of Christianity Today* 13 (2).

Pust, Joel. 2019. Intuition. *The Stanford Encyclopedia of Philosophy* (Summer 2019 Edition). Edward N. Zalta (ed.). Available at https://plato.stanford.edu/archives/sum2019/entries/intuition/.

Ratzsch, Del. 2001. *Nature, Design and Science: The Status of Design in Natural Science*. Albany, NY: State University of New York Press.

Ratzsch, Del, and Koperski, Jeffrey. 2019. Teleological Arguments for God's Existence. *The Stanford Encyclopedia of Philosophy* (Summer 2019 Edition). Ed. Edward N. Zalta. Available at https://plato.stanford.edu/archives/sum2019/entries/teleological-arguments/.

Reiss, John. 2009. *Not by Design: Retiring Darwin's Watchmaker*. Berkeley, CA: University of California Press.

Ruse, Michael. 2003. *Darwin and Design: Does Evolution Have a Purpose?* Cambridge, MA: Harvard University Press.

———. 2004. The Argument from Design: A Brief History. In *Debating Design: From Darwin to DNA*, ed. William A. Dembski and Michael Ruse, 13–31. Cambridge: Cambridge University Press.

Sagan, Carl. 1985. *Contact*. London: Orbit.

Salmon, Wesley. 1990. *Four Decades of Scientific Explanation*. Minneapolis, MN: University of Minnesota Press.

Schupbach, Jonah. 2005. Paley's Inductive Inference to Design. A Response to Graham Oppy. *Philosophia Christi* 7 (2): 491–502.

Sehon, Scott. 2005. *Teleological Realism: Mind, Agency and Explanation*. Cambridge, MA: MIT Press.

Shostak, Seth. 2005. SETI and Intelligent Design. Space.com. December 1. Available at http://www.space.com/1826-seti-intelligent-design.html.

Sober, Elliott. 2000. *Philosophy of Biology.* (2nd ed.). Dimensions of Philosophy Series. Boulder, CO: Westview Press, pp. 32–33.

———. 2003. The Design Argument. In *God and Design: The Teleological Argument and Modern Science*, ed. Neil A. Manson, 27–54. New York: Routledge.

Stenger, Victor J. 2011. *The Fallacy of Fine-Tuning: Why the Universe Is Not Designed for Us.* Amherst, NY: Prometheus.

Susskind, Leonard. 2006. *The Cosmic Landscape: String Theory and the Illusion of Intelligent Design.* New York, NY: Back Bay Books.

Swinburne, Richard. 2004a. *The Existence of God.* Oxford: Oxford University Press.

———. 2004b. The Argument From Nature Reassessed. In *Debating Design: From Darwin to DNA*, ed. Dembski & Ruse, 298–310. Cambridge: Cambridge University Press.

———. 2010. God as the Simplest Explanation of the Universe. *European Journal for Philosophy of Religion* 2 (1): 1–24.

———. 2012. Why Hume and Kant Were Mistaken in Rejecting Natural Theology. In *Gottesbe-weise als Herausforderung für die Moderne Vernunft*, ed. Th. Buchheim, F. Hermanni, A. Hutter, and Ch. Schwöbel, 317–334. Tübingen: Mohr Siebeck.

Taliaferro, Charles. 2013. Philosophical Critique of Natural Theology. In *The Oxford Handbook of Natural Theology*, ed. Russell Re Manning, 385–394. Oxford: Oxford University Press.

Turner, Denys. 2004. *Faith, Reason and the Existence of God.* Cambridge: Cambridge University Press.

Turner, Scott. 2017. *Purpose and Desire: What Makes Something "Alive" and Why Modern Darwinism has Failed to Explain It.* San Francisco, CA: HarperOne.

Walsh, Denis. 2008. Teleology. In *The Oxford Handbook of Philosophy of Biology*, ed. Michael Ruse, 113–137. Oxford: Oxford University Press.

Wilkins, John, and Wesley Elsberry. 2001. The Advantages of Theft Over Toil: The Design Inference and Arguing from Ignorance. *Biology and Philosophy* 16 (5): 709–722.

Williamson, Timothy. 2000. *Knowledge and Its Limits.* Oxford: Oxford University Press.

Woodward, James. 2003. *Making Things Happen: A Theory of Causal Explanation.* Oxford: Oxford University Press.

———. 2017. Scientific Explanation. *The Stanford Encyclopedia of Philosophy* (Fall 2017 Edition). Edward N. Zalta (ed.). Available at https://plato.stanford.edu/archives/fall2017/entries/scientific-explanation/.

Ylikoski, Petri. 2001. *Understanding Interests and Causal Explanations.* Helsinki: University of Helsinki.

CHAPTER 4

Not by Selection Alone: Evolutionary Explanations and Their Requirements

Statements of the fundamental conflict of evolution and design-based explanations are not difficult to find in the literature, even among biologists who otherwise have a positive attitude toward religious belief. For instance, according to Francisco Ayala:

> The theory of evolution conveys chance and necessity jointly intricated in the stuff of life; randomness and determinism interlocked in a natural process that has spurted the most complex, diverse, and beautiful entities in the universe: the organisms that populate the earth, including humans who think and love, endowed with free will and creative powers, and able to analyze the process of evolution itself that brought them into existence. This is Darwin's fundamental discovery, that there is a process that is creative though not conscious, a process that creates design without necessitating a Designer. (Ayala 2010, 843)

In this typical account, the basic assumption is that evolution and design are explanations for the same evidence, and thus the triumph of evolutionary biology makes design unnecessary (though the existence of a designer is still typically held to be compatible with evolution). Most contemporary defenders of biological design arguments, such as those in the ID movement, have accepted this basic assumption, and consequently their defense of biological design arguments requires criticizing the

E. V. R. Kojonen, *The Compatibility of Evolution and Design*, Palgrave Frontiers in Philosophy of Religion, https://doi.org/10.1007/978-3-030-69683-2_4

plausibility of evolutionary explanations. In contrast, critics of biological design arguments contend that the evidence supports evolution, and thus refutes design. The opposition can be traced back to Darwin's own views and his discussions with Asa Gray (1888b [1874]), who believed in the compatibility of the biological design argument and evolution. As noted, according to Darwin (1958 [1887], 87–88), "The old argument of design in nature, as given by Paley, which formerly seemed to me so conclusive, fails, now that the law of natural selection has been discovered."

In the next chapter, I will question the assumption of conflict in more detail. But before that, much more needs to be said about evolutionary explanations themselves. The appeal of design arguments arises partly from the intuition that the apparent teleology of biology is difficult to produce by any undirected process, and perhaps even impossible to explain without intelligent guidance. This book assumes that the critiques presented by the ID movement fail, and that evolutionary explanations are plausible. Nevertheless, as Gray's example shows, it is a separate matter to conclude that evolution was not difficult, or did not require design even to get started. As de Duve (1984) pointed out, evolutionary chance and natural selection "did not operate in a vacuum. It operated in a universe governed by orderly laws and made of matter endowed with specific properties. These laws and properties are the constraints that shape the evolutionary roulette and restrict the numbers that it can turn up."[1] The decades after de Duve's statement have greatly strengthened this conclusion. I will begin with the basics, and then go on to argue that both recent biology and the responses to the ID movement show that evolution indeed has very demanding preconditions.

4.1 The Darwinian Blind (but Powerful) Watchmaker

The Basics of Evolutionary Explanations

The traditional three basic pillars of the evolutionary mechanism are variation, inheritance, and differential reproduction or survival (Bowler 2009; Ridley 2004; Freeman and Herron 2007; Sober 2008). While Darwin did not have a developed theory about the origin of variation, it was clear that organisms of the same species varied in their characteristics. Artificial breeding demonstrated the inheritability and accumulation of variations, resulting in new breeds of pigeons and dogs. Darwin's insight was to realize that a similar process of selection could also be working in nature. Some naturally occurring variations could be more or less beneficial to the species, thus leading to different rates of survival and reproduction. In the struggle of life, the forms more adapted to the environment would eventually succeed over less adapted forms, and new variations could build on older ones, resulting in evolution over time.

Darwin (2008 [1859], 84) presented natural selection as an almost omniscient agent: "It may be said that natural selection is daily and hourly scrutinizing, throughout the world, every variation, even the slightest; rejecting that which is bad, preserving and adding up all that is good." Selection remains a central part of most current understandings of evolution, as well. However, later, with the advent of population genetics and neutral evolutionary theory, it was emphasized that natural selection is actually blind to most variation, since most variation will have statistically insignificant effects on survival. There is also debate over whether large-scale evolution might be affected by different mechanisms in addition to those affecting small-scale evolution (Sepkoski 2008; Lewens 2007).

I should note at this point that there have recently been calls for substantial expansion and modification of evolutionary theory (Laland et al. 2014; Pigliucci and Müller eds. 2010). For instance, evo-devo (the synthesis of evolutionary biology and developmental biology) has provided new ideas about the interplay between different genes and the organismal form, and the discussion of epigenetics has highlighted the existence of heritable variation outside of genes. The study of factors like sexual selection and niche construction has emphasized the role of biological organisms in shaping their own evolution. Thus, evolutionary biology has moved away from the idea that all features of form must be explained in virtue of their function, caused through natural selection working on random mutations. However, this does not mean that adaptive complexity does not need to be explained. And the standard Darwinian mechanism of mutation and selection is still commonly seen as essential for explaining complex biological adaptations even by many of those most critical of standard theory (Fodor and Piattello-Palmarni 2011).[2] Moreover, the standard account is also the one that is typically thought to be the most opposed to any ideas of design; thus, if design turns out to be compatible with it then it will surely also be compatible with the extended synthesis. For these reasons I will begin here with the standard account.

As noted, Darwin himself argued that knowledge of evolution removes the need for appealing to direct divine design to explain biological adaptations. He did not even agree with those who, like the American evolutionist Asa Gray (1810–1888), believed that God directed the evolutionary process by directing the variations which were then passed on to the next generation. For Darwin, writing to Lyell, "The view that each variation has been providentially arranged seems to me to make Natural Selection entirely superfluous, and indeed takes the whole case of the appearance of new species out of the range of science. [...] Gray's notion seems to me to smash the whole affair" (Darwin 1861). If the variations were indeed divinely guided, and this supposition was required to explain biological adaptations, then this would have, for Darwin, made natural selection of secondary importance in explaining these adaptations. Darwin emphasized, instead, that each variation should be understood as random, as just one out of thousands and millions of such changes, out of which natural selection sifts the beneficial and eliminates the harmful ones.

Gray (1860) had argued that divine design might be necessary for evolution in two main ways: (1) determining the emergence of positive variations, which was as yet unexplained by the science of Darwin's day, and (2) setting up the process, by providing suitable laws and an environment. According to Gray, it makes no difference whether the ordered complexity of life is produced through a secondary cause, such as evolution, or through a miraculous act of creation that exceeds the normal operation of nature. In either case, he claims, the design can only be inferred from the purposeful arrangement of the end result. Biological science is the same for the theist and the atheist, just as physics is the same—instead, Gray argues, "the difference appears only when the inquiry is carried up to the question of primary cause—a question which belongs to philosophy" (Gray 1860).

However, though Darwin emphasized the compatibility of evolutionary theory with theism, he nevertheless rejected Gray's harmonization of the biological design argument and evolution (Browne 2010; Johnson 2015). Darwin's most lengthy commentary on Gray's argument can be found in the last pages of his *Variation of Animals and Plants Under Domestication* (Darwin 1868). In the passage, Darwin emphasizes the importance of selection as the real explanation of biological features, over against variation. He does this through the analogy of an architect building from a random assembly of stones. Due to the great influence of

Darwin's interpretation of his own theory, the analogy is worth quoting at length:

> If an architect were to rear a noble and commodious edifice, without the use of cut stone, by selecting from the fragments at the base of a precipice wedge-formed stones for his arches, elongated stones for his lintels, and flat stones for his roof, we should admire his skill and regard him as the paramount power. Now, the fragments of stone, though indispensable to the architect, bear to the edifice built by him the same relation which the fluctuating variations of organic beings bear to the varied and admirable structures ultimately acquired by their modified descendants. […] Can it be reasonably maintained that the Creator intentionally ordered, if we use the words in any ordinary sense, that certain fragments of rock should assume certain shapes so that the builder might erect his edifice?
>
> If the various laws which have determined the shape of each fragment were not predetermined for the builder's sake, can it be maintained with any greater probability that He specially ordained for the sake of the breeder each of the innumerable variations in our domestic animals and plants;—many of these variations being of no service to man, and not beneficial, far more often injurious, to the creatures themselves? […] But if we give up the principle in one case,—if we do not admit that the variations of the primeval dog were intentionally guided in order that the greyhound, for instance, that perfect image of symmetry and vigour, might be formed,—no shadow of reason can be assigned for the belief that variations, alike in nature and the result of the same general laws, which have been the groundwork through natural selection of the formation of the most perfectly adapted animals in the world, man included, were intentionally and specially guided. (Darwin 1868, 426–428)

Darwin argued that the shape of the stones was random in the sense of being unrelated to the usefulness of the stones to the architect in this example. Although deterministic causes could in principle be found for the shape of each stone—and hence they are not random in the absolute sense—it is nevertheless, for Darwin, incredible to think that the shape of each rock was intended by the Creator to be used in the architect's building. Darwin claims that it is just as incredible to think that the mutations in animals are ordained by God in order to facilitate evolution—particularly if the harmful nature of many mutations is taken into account. But if we give up the existence of divine guidance of variation even in one case, even in the case of the selective breeding of pigeons, then it seems—says

Darwin—arbitrary to say that divine guidance was necessarily involved even in mutations used by natural selection.

In his theory of evolution, Darwin emphasizes, the variations are unrelated to the needs of the organisms or evolution. It is selection that does the building out of an enormous amount of variation over long periods of time. This, for Darwin, emphasizes the explanatory importance of selection and the unguided nature of variation. He further emphasizes that nothing in this explanation requires or implies that the variations are directed. Indeed, if each variation were preordained from the beginning, then natural selection is in fact unnecessary as an explanation.

On the face of it, all of this seems aimed merely at Gray's argument for the possible divine guidance of mutations. But it seems that an argument against the explanatory relevance of the primary cause is also included in the analogy of the architect. In the analogy, it is the architect who is primarily responsible for the organization of the house, even though of course the house could not be constructed without the existence of the laws and processes of nature which provide the building blocks. Similarly, it seems that for Darwin, natural selection in contingent historical circumstances is the primary factor explaining animal form, leaving little role for divine design as the cause of form. As a result, to return to Darwin's previously quoted statement in his *Autobiography*, "There seems to be no more design in the variability of organic beings and in the action of natural selection, than in the course which the wind blows." In Darwin's mature understanding, then, selection itself seems to explain the orderliness of biological nature, so that explaining biology's order no longer necessarily requires any other "primary cause" to have intentionally caused it. It seems to me that this is why, for Darwin, Paley's design argument now fails to be convincing.[3]

Darwin's analogy also highlights evolution's dependency on deep time, time giving the opportunity for the countless chance variations that natural selection then works on. The advancement of geology meant that evolution did have an incomprehensibly long amount of time to sort through variations and arrive at functional forms. But the history of the cosmos is still finite, and so are the possibilities of natural processes. Thus, as much as deep time helps, it could in principle have been the case that even vast amounts of time would not be sufficient for the even more vast problems that evolution had to overcome. Indeed, evolutionary theory is built on the assumption that chance alone is insufficient as an explanation. It is only because of the limitations of time and resources that natural selection is even needed. As the famous thought experiment about infinite monkeys writing on an infinite number of typewriters seems to show, with infinite amounts of time and trials even chance events might eventually suffice to produce complex and interesting outcomes.[4]

Many contemporary evolutionists continue to emphasize the importance of selection and contingent historical factors when they argue that evolutionary explanations make design unnecessary. Daniel Dennett, for example, focuses on evolution's ability to split up the difficulty of the evolutionary process into small pieces, none of which is difficult to explain as an undesigned chance event.

> Paley was right in saying not just that Design was a wonderful thing to explain, but also that Design took Intelligence. All he missed—and Darwin provided—was the idea that this Intelligence could be broken up into bits so tiny and stupid that they didn't count as intelligence at all, and then distributed through space and time in a gigantic connected network of algorithmic process. (Dennett 1995a, 133)

Dennett argues that because each individual evolutionary event can be plausibly explained as the working of natural selection on chance mutations, rather than design, the entire series of evolutionary events across space and time should also be inferred to be without design. What is true of the parts is, for Dennett, here also true of the whole.

However, it seems to me that there is still more to be said for the dependency of evolution on the overall order of the cosmos. As noted by the Thomist natural theologians discussed in the previous chapter, selection itself is a result of the environment, and thus not an independent actor in the same way as the architect in Darwin's analogy. Moreover, the relationship of parts and wholes in evolution is not necessarily as simple as Dennett's argument seems to imply. Even in the case of human design, one might argue that individual elements of a building designed by humans could be explained as accidental. It is only when we consider the whole of the building that the necessity of design becomes apparent. The same could well be true regarding evolution. For instance, it is not clear whether the direction taken by the evolutionary series as a whole is impacted by other factors beyond those that we can discern in an isolated evolutionary event. The possibility of accumulating these mutations and forming new forms of life might still require other factors, such as the existence of an environment that supports the existence of this kind of form, the existence of natural selection, and the existence of a functional series of intermediate steps that connect functional forms. If such preconditions play a substantial role in making evolution possible, then this also needs to be taken into account in constructing the picture of evolution as a whole, and how it relates to design arguments. Supposing that these preconditions are the result of design, then it would no longer be true that evolution proceeds without design.

André Ariew gives us another example of how evolutionary explanations could show the unnecessity of design. Like Darwin, Ariew argues that evolution can be explained without any overarching purpose, simply by reference to the creatures' needs to pursue their own survival and reproduction—resulting in competition for limited resources, and thus natural selection. Thus, according to Ariew (2007, 172), adaptations "are the result of the collective activity of individuals that are simply striving to survive and reproduce on their own, without a mind to the collective effect." As an analogy, Ariew compares evolution to economics—a commonly used analogy. Referencing Adam Smith's famous "invisible hand" that produces benefits for all from the selfish behavior of the few, Ariew claims that no overarching purpose is required to explain economic development, and no overarching purpose is similarly required to explain biological evolution (similarly Shanks 2007, 45). The credibility of the analogy with capitalism is somewhat context-dependent. Since the 2007–2009 financial crisis, there is broader recognition of the dependency of capitalist systems on at least some planning, regulation, security, and infrastructure as preconditions for the benefits of a free market system. Financial profits can indeed motivate people to plan, produce, and provide services that meet real needs, provide work and a livelihood, and thus improve the quality of life. However, unregulated, the pursuit of financial profits directed by greed can also lead to the exploitation of workers (including child labor and slave labor), the selling of harmful products, and the destruction of the environment. Smith himself recognized the necessity of regulation, arguing this extensively in the case of banks (Rockoff 2011). Indeed, for readers in Smith's day the idea of an invisible guiding hand would have had providential connotations (Oslington 2012).

Thus the analogy of economics could also be understood to support Gray's point about the necessity of design. Understood in its original providential context, the idea of the "invisible hand" guiding economic growth leads naturally to the idea that evolution might also be based on a providential order that makes it possible. In a capitalist economy, the profit-seeking actions of individuals can only result in the increase of common good because of the fulfillment of certain preconditions, and the existence of regulation and/or virtue. Following Gray, one might argue that it is quite serendipitous, even suggestive of providence, that under the conditions that exist in nature, the behaviour of creatures pursuing their own survival just happens to result in evolution. We could then further ask if this fortuitous link between creatures' survival and evolution happens by chance or by design—Gray (1860), for instance, thought that the case for design is very strong simply based on the end result, with the discovery of the evolutionary process not reducing the strength of the case.

There is a need to delve more deeply into these questions, and the problem of explaining complex adaptations provides a good initial test case. Complex structures like the human eye have historically provided the most commonly referenced evidence of design in biology, and demonstration of the possibility of evolving such structures has thus been seen as a necessary part of refuting the design argument. As noted, in this book I am assuming the success of Darwinian explanations, and use this to ask whether the unnecessity of divine interventions in the history of life then refutes the biological design argument. I have invited readers who disagree with evolutionary explanations to also adopt this assumption for the sake of the argument—for I think that the question of the in principle compatibility of evolution and design arguments would be philosophically interesting, even if evolutionary explanations were overturned. However, granting the success of evolutionary explanations still does not mean that the problem of explaining complex adaptations is irrelevant to my argument. In practice, the best defenses of the power of the evolutionary mechanism have been written in response to skeptics, just as many of the tenets of Christian doctrine were sharpened in reaction to heresies. Analyzing these responses will let me probe deeper into just how evolution might conflict with the design argument.

Explaining Complex Adaptations

Darwin recognized the explanation of complex adaptations like the eye as a potential difficulty for his theory. Paley had used the eye as a clear example of design, and contemporary anti-evolutionists (e.g., Berlinski 2004) continue to refer to it as a problem for evolution. What use, they ask, would just a part of an eye have, when the evolution of vision would surely require many simultaneous changes, not only to produce the eye, but also the required brain tissue? How could the steps leading up to an eye be useful from the point of view of natural selection? However, Darwin (2008 [1859], chapter VI) concluded that "if it could be demonstrated that any complex organ existed, which could not possibly have been formed by numerous, successive, slight modifications, my theory would absolutely break down. But I can find out no such case." The Darwinian argument is that because we have evidence for the actuality of evolution from a wide array of facts, and because we can observe variation and selection working in practice, this gives grounds for believing that complex organs can also be explained by the same stepwise process. In many cases, viable evolutionary steps can also be proposed. It is acknowledged that we do sometimes have considerable difficulty in conceiving of evolutionary explanations for particular organs. Denis Noble (2008, 109), no creationist, compares the problem to the modifying and improving of an aircraft while it is in flight and all systems are in use. However, insofar as a process of gradual evolution can already be traced for many complex organs, this lends credibility to the claim that other complex organs can also be similarly explained. As a human enterprise, evolutionary science will always be based on incomplete evidence of the complex past, and as such we should not be surprised that evolutionary explanations are also incomplete. But this does not in itself prevent them from being good explanations.

Dawkins' defense of the possibility of evolution in his *The Blind Watchmaker* (1986) continues to be referenced even in biology textbooks, and also forms the starting point for the ID movement's critique of evolutionary explanations, so I will now consider his account. While Dawkins admits that complex biological adaptations like the human eye appear purposefully designed, he nevertheless argues that mutation and selection can in fact explain such structures. Consider the eye, the example that Paley (2008 [1802], chapter VI) thought was so strong that "were there no example in the world, of contrivance, except that of the *eye*, it would be alone sufficient to support the conclusion which we draw from it, as to the necessity of an intelligent Creator." Dawkins agrees that it is too implausible to assume that the human eye emerged simply by chance, in a single step. But he does think that it is plausible for such complex organisms to have evolved via a "blind watchmaker" process that requires no purposeful oversight, consciousness, or knowledge of the future. He presents the basic preconditions of such evolution as a response to five questions:

1. Could the human eye have arisen directly from no eye at all, in a single step?
2. Could the human eye have arisen directly from something slightly different from itself, something that we may call X?
3. Is there a continuous series of Xs connecting the modern human eye to a state with no eye at all?
4. Considering each member of the series of hypothetical Xs connecting the human eye to no eye at all, is it plausible that every one of them was made available by random mutation of its predecessor?
5. Considering each member of the series of Xs connecting the human eye to no eye at all, is it plausible that every one of them worked sufficiently well that it assisted the survival and production of the animals concerned? (Dawkins 1986, 77–81)

Following Darwin, Dawkins argues that it is not that difficult to imagine that the human eye could have emerged from a slightly simpler eye that would still have been useful to possess, and thus its evolution would have been favored by natural selection. And then we can imagine that this eye could have been evolved from a slightly simpler eye, and so on, hopefully until we end up at just a simple light-sensitive spot, which we know is still useful from studying various living organisms. This just requires a continuous series of possible and biologically viable forms from start to finish (Nilsson and Pelger 1994, compare Berlinski 2004).

The dependency of evolution on the existence of such pathways is widely acknowledged. As Jerry Coyne (2007, 42) also writes, "It is indeed true that natural selection cannot build any feature in which intermediate steps do not confer a net benefit on the organism." However, the argument following from this admission is that the existence of any feature which could not have evolved by such intermediate steps has not been demonstrated. Instead, Coyne argues that plausible evolutionary explanations have indeed been developed for many complex systems.[5]

As further evidence, Dawkins (1986, 66–68) turns to a digital illustration of the evolutionary mechanism's power in producing a meaningful sentence, the Shakespearean "METHINKS IT IS LIKE A WEASEL."[6] This sequence of 28 letters is far too long to generate randomly in a reasonable timeframe, but Dawkins explains that his algorithm allows one to insert any random sequence of 28 letters, which will then be mutated step by step into the target sequence. Dawkins' algorithm treats the sequence like an organism that is mutated slightly by changing letters in each "generation," and the fittest sequences (those most closely resembling the target sequence) are then selected to be mutated again. With the repetition of the procedure, the digital organisms quickly converge on the target sequence. Dawkins' point is that, here, random mutation (the variability in the letters that make up the randomly generated sentence) together with heritability and natural selection (which ensure that correct letters are passed on to the next generation) indeed work in producing complex sequences that would be far too improbable to generate by chance.

These kinds of algorithms are called evolutionary algorithms and genetic algorithms, and Dawkins' simple experiment has since been repeated with far more impressive programs, such as *Avida*, *Tierra*, and *Ev*. For example, the Avida simulation creates "digital organisms," which self-replicate, compete, and evolve further functions beyond the functions possessed by the original "organisms." Avida was set up to allow for the evolution of complex functions by also favoring the presence of simpler functions which were stepping stones toward that complex function. The authors argue that "these findings show how complex functions can originate by random mutation and natural selection" (Lenski et al. 2003, 139). However, corroborating Dawkins on the presuppositions of the evolutionary process, "our experiments showed that the complex feature never evolved when simpler functions were not rewarded" (Lenski et al. 2003, 143). Indeed, the authors had to be quite careful about defining such rewards in order for the simulation to produce evolution as they intended.

Evolutionary algorithms have even been successfully used to come up with new solutions to real-world problems. Possibly the most celebrated example of this is the evolutionary design of a new type of antenna by NASA engineers. The antenna has a very irregular shape that was arrived at by an evolutionary algorithm that simulated different antenna shapes and tested them against parameters taken from real-world physics, modifying an antenna design by trial and error. The final design was unanticipated by the engineers, but proved useful in practice, and has been put to use (Venkataraylu and Ray 2003; John et al. 2004). Algorithms like this work using simple principles of trial and error, guided by the parameters inputted into the program that work as selection criteria (Simon 2013). Nevertheless, once the basic parameters of the program are in place, the programs can then use evolutionary principles to achieve results that are similar to direct design, and indeed could be mistakenly inferred to be the result of direct design by people who were unaware of the use of the algorithms. Thus, in the debate over evolutionary explanations, genetic algorithms have commonly been used to argue that the basic mechanisms of evolution are powerful enough to produce complex, purposeful order without the need for the intervention of a designer.

Why should the evolutionary mechanism then refute the need for design as an explanation of biological adaptations? At least in these genetic algorithms, design really is an important component of explaining the resulting pattern, since the whole functioning of the genetic algorithm would be impossible without design. What this reveals is that in standard accounts, design is typically understood to mean the direct intervention of the designer to produce the result in question, rather than any presence of design behind the system. When evolution shows the unnecessity of interventions, it therefore also undercuts design arguments. In this understanding, design arguments function by comparing chance and design as explanations for biology. Darwinism, however, shows that chance and design are not the only alternative explanations. Rather, we can construct an explanation combining natural selection and chance processes.

It is interesting to note already at this point that the case of genetic algorithms—like the earlier example of economics—is a case in which the possibility of evolution depends on design. From the point of view of this book, it is interesting how difficult it has proven to be to present an analogy of evolution that does not depend on design. The personification of natural selection is also commonplace in the literature, from Darwin onwards (Riskin 2016; Livingstone 2020). However, these are analogies. In order to get beyond overly simplistic portrayals of the process of natural selection and mutation, there is a need to further analyze the preconditions of the evolutionary process. I will begin with responses to three main anti-evolutionistic arguments. In all cases, I will argue that the defensibility of evolutionary explanations depends on certain preconditions of the evolutionary process being fulfilled. This, in turn, contributes to the saving of the biological design argument.

4.2 Answering Critiques of Evolutionary Explanations

The Conservation of Information

Dembski's design argument comes close to the inference to the only explanation described in the previous chapter. His eliminative design argument is based on attempting to eliminate "chance and necessity," meaning coincidences, probabilistic natural processes, and law-like regularities as sufficient explanations for some phenomena in nature. For this purpose, Dembski (1998, 2002a) formulates the criterion of "specified complexity." For Dembski, a short sequence of letters like "CAT" is specified but not complex—finding such a pattern written in Scrabble letters might still be explained as the coincidental result of pieces dropped randomly. A long sequence of random letters is complex but not specified—it does not match a previously known pattern. A Shakespearean wording like METHINKS IT IS LIKE A WEASEL, on the other hand, is both complex and specified. This type of specified complexity is, for Dembski, a reliable criterion for the existence of an intelligent designer, and we can reliably infer that chance and necessity do not explain it.

There is a vast literature responding critically to Dembski's design argument and his criterion of specified complexity, and several overviews are available (Loikkanen 2019; Berhow 2019; Perakh 2004). For example, it seems that his argument is overly focused on eliminating chance and necessity as explanations, rather than providing positive reasons for thinking that design is a good explanation for specified complexity—though Dembski does also attempt to provide positive reasons for belief in design as the correct explanation (Himma 2005; Dawes 2007). But I have already considered the logic of the design argument in the previous chapter. What concerns me here is evolution as a counterargument, and how Dembski responds to it. Before Dembski, Dawkins himself formulates a very similar criterion in *The Blind Watchmaker*. He writes:

> Hitting upon the lucky number that opens the bank's safe is the equivalent [...] of hurling scrap metal around at random and happening to assemble a Boeing 747. Of all the millions of unique and, with hindsight, equally improbable, positions of the combination lock, only one opens the lock. Similarly, of all the millions of unique and, with hindsight, equally improbable, arrangement of a heap of junk, only one (or very few) will fly. The uniqueness of the arrangement that flies, or that opens the safe, is nothing to do with hindsight. It is specified in advance. (Dawkins 1986, 8)[7]

Here, just like Dembski, Dawkins argues that a "specified" complex pattern cannot be plausibly explained as just an improbable coincidence. This is illustrated by Dawkins' previously described algorithm, which is able to generate the words METHINKS IT IS LIKE A WEASEL through a stepwise process mimicking evolution. But Dawkins goes on to argue that evolution can solve the problem of originating biological specified complexity, because evolution can build it in small bits rather than having to generate it all at once. Thus, while chance alone cannot explain specified complexity, the combination of chance and necessity can (similarly Perakh 2004).

Against this, Dembski argues that there is a general law of "conservation of information" which prevents even the combination of chance and necessity, as in evolution, from explaining the origination of new specified complexity without intelligence. Dembski and his colleague Robert J. Marks II (2009a, b) define the law as the statement that any search process "must build into the search at least as much information as the search displays in raising the probability of success." As Dembski and Marks note, understanding evolution as a search process working by trial and error is common also outside the literature of the ID movement (e.g., Dennett 1995a; Kauffmann 2000). However, if evolution is a search process, then it seems that common mathematical results that have been proven in the case of computerized search algorithms should also apply to evolution. For example, the "No Free Lunch" (NFL) theorem, first stated by Wolpert and Macready (1997), claims that there is no universal search strategy that would perform better than a random search when averaged across all possible problems. In order to be able to formulate a search strategy that works better than a random search, one needs to formulate the strategy to match the problem and the search target. And this, Dembski argues, always requires design—even in the case of the evolutionary search.

Dembski and Marks' thesis seems credible in the case of Dawkins' (1986) WEASEL algorithm. Here the target phrase "METHINKS IT IS LIKE A WEASEL" is already built into the program at the beginning, and is used as the selection criterion for the fitness of each variant. This program uses in-built information to direct the process of mutation and selection to eventually converge on the correct order of letters. The process is teleological, and the end result can be understood to have been inevitable once we understand how the algorithm works. Instead of being an example of how information can arise without a designer, the algorithm does not generate any new specified pattern that was not already built into the program at the start.

The same result seems more difficult to demonstrate in terms of evolutionary simulations like Avida, and especially in cases where evolutionary algorithms have produced new blueprints that were wholly unanticipated by the programmers, such as the NASA antenna. However, Dembski and his colleagues argue that also in these cases, a working algorithm requires a great deal of built-in information, at least as much as it produces (Dembski et al. 2016; Ewert et al. 2009; Ewert 2016, 2017). In the case of Avida, as the authors of the program themselves note (Lenski et al. 2003, 143), the possibility of evolving complex functions was only realized after the program was edited to reward the evolution of simpler functions that formed the building blocks of the more complex functions. In the absence of such a reward system, complex functions failed to evolve. Against the potential complaint that the authors "stacked the deck" in rewarding the evolution of simple functions that then formed parts of more complex functions, the Avida authors do not deny stacking the deck, but retort that this was necessary in order to simulate evolution. After all, in evolution, as well, complex functions are built only on top of simpler ones, where a clear pathway exists for building the more complex function.

However, according to Dembski et al. (2016, 3.6), this means that the outcome is in fact dependent on design, and that Avida is teleological. They claim that the situation is similar in the case of the antenna-building algorithm, the program being directed to simulate and evaluate for a good antenna arrangement, rather than something else. Their overall conclusion is that each genetic algorithm only works to solve a specific problem, and only because it is guided by the parameters built into it by the programmers. However, the capacities of the programs have clear limitations: "An evolutionary program written to design an antenna will never develop the ability to play chess. And a fixed environment to guide a stochastic process into evolving complexity on the scale of man is unimaginably improbable" (Dembski et al., chapter 4).

But how convincing is the argument? In response to the critique of evolutionary algorithms, Roy Meester has argued that evolutionary algorithms should indeed no longer be used as examples of the power of the evolutionary mechanism, arguing that "the discussion around the NFL theorem demonstrates that in a computer simulation which constitutes a search algorithm, a 'special' target can only be reached when you carefully choose your algorithm around the target, that is, if the algorithm is intrinsically teleological" (Meester 2009, 470). However, against Meester, I am inclined to agree with the assessment of Boudry, Blancke and Braeckman (2011), who argue that evolutionary algorithms are a valid way of simulating the principles of evolution, and that they must include such guiding information in order to simulate natural selection. In Dawkins' WEASEL algorithm, for example, it is not inappropriate for there to be a selecting condition for the letters, because there are also selective conditions in natural environments. The point of the algorithms is to demonstrate the possibility of building complex order by a stepwise process involving differential survival, if the preconditions of such an evolutionary process are met.

The more pressing criticism is that the No Free Lunch theorem (and other similar results) do not invalidate the theory of biological evolution. Yes, mathematician Olle Häggström (2007a, 2007b) admits, NFL theorems are valid and there is no difference between a blind search and other search processes when the search space that is investigated is randomly arranged. However, there is no guarantee that the search space that biological evolution searches through is random in this way, and hence there is no necessary obstacle to evolution. Rather, functional forms can be situated close to each other, making them accessible to a stepwise evolutionary search, just as in the genetic algorithms. As Häggström (2007a, 227) writes, "it is to a large extent clustering properties that make local search algorithms [...] work better than, say, blind search." Since nothing about the NFL theorem can show that biological search space is not in fact arranged in a way that allows for evolution, it cannot show that evolution does not work. Indeed, says Häggström, the basic fact that forms are variable at all already suggests that functional forms are not randomly distributed across the search space. Against Dembski, he argues that we cannot just assume that the biological search space is not arranged in a way that in fact allows for evolution, but would have to demonstrate it.[8]

It is unclear whether Dembski and his colleagues are right about the existence of a general law of conservation of information and its applicability to genetic algorithms, so that no new information is indeed created even in cases like the building of the antenna. However, this does leave valuable points mostly uncontested by critics. For instance, it does seem credible to conclude that in genetic algorithms, the production of new information does depend on the existence of prior information in the system, and that such algorithms can only solve problems that they are specifically designed to solve. The systems must be built to reward stepwise growth, and the selective conditions must be adapted to the problem being solved. Thus, in these simulations, the possibility of evolution depends on design, demonstrating that there is no necessary contradiction. One can further argue that due to the strong dependency of the algorithm on design, the products of the simulated evolutionary process are revelatory of the intelligence of the programmer.

Indeed, Dembski and colleagues have argued that the critics have not refuted their point about design. The basic argument, they say, is against the idea of Darwinian evolution as an unguided process, not dependent on intelligence, rather than against evolution as such. Already in *No Free Lunch*, Dembski (2002a, 212) argues that "even if Darwinian evolution is the means by which the panoply of life on earth came to be, the underlying fitness function that constrains biological evolution would not be a free lunch and not a brute given, but a finely crafted assemblage of smooth gradients that presupposes much prior specified complexity." In a later paper, Dembski and Marks (2009b) similarly argue that "when we turn to evolutionary biology, we find clear evidence of teleology: despite Dawkins's denials, biological evolution is locating targets. Indeed, function and viability determine evolution's targets [...] and evolution seems to be doing a terrific job finding them." And Ewert (2016) clarifies: "We are not claiming and have not claimed that evolution simply doesn't work in any environment. We have claimed that it only works in certain environments. This is the entire point behind the concept of active information and conservation of information: if you have an environment where evolution is successful it must be due to the configuration of the environment."

The conclusion that the argument aims at, then, is that the information in biology ultimately points to design, because unguided evolutionary processes cannot explain it starting from nothing, but always depend on the existence of prior information, which can ultimately only be explained by reference to design. Häggström (2007b), commenting on this, notes that design is indeed a possible explanation for the orderliness of the cosmos that makes evolution possible. However, he argues that this takes the discussion into deep metaphysical and even theological waters, and is irrelevant for the viability of evolutionary explanations within biology. But from my perspective, the moving of the question of design to the realm of metaphysics and theology, and stepping away from the idea of an explanatory competition between evolution and design, would be a positive development. This would contribute to the case that, at least to some extent, evolution pushes back the problem of the origin of biological teleology to the conditions that make evolution possible.

While the aforementioned ID proponents argue the case for the dependency of Darwinian evolution on design, the mainstream of the ID movement at the same time nevertheless rejects the possibility of theistic evolution as inconsistent with design arguments (Moreland et al. ed. 2017; Kojonen 2016, chapter 10). Perhaps this is partly because Dembski and his colleagues do not prefer to argue for design by focusing on the preconditions of the evolutionary process. Rather, Dembski claims that while pathways that enable evolution exist in genetic algorithms, they do not exist in nature:

> Granted, I don't think this is the strongest place to argue for biological design. The strongest place is to argue head-on that the fitness landscapes are not smooth. But design at the level of fitness landscapes (under the supposition that Darwinism is correct) remains to my mind a powerful argument in light of all the cases we know where the Darwinian mechanism operates but fails to produce anything interesting. (Dembski 2002b)

To demonstrate the non-viability of evolutionary processes in nature, Dembski (2002a, chapter 5) turns to other arguments beyond his proposed law of conservation of information, chiefly to Behe's irreducible complexity. As it turns out, the discussion surrounding Behe's argument is also of interest in demonstrating how tight the conditions for the possibility of evolution are.

Irreducible Complexity

Many proponents of ID have taken Dawkins' "blind watchmaker" as the main defense of evolutionary biology that they need to respond to, and Behe also references many facets of Dawkins' argument. For instance, Behe (2006, 38) argues that the "simple" light-sensitive spot in Dawkins' account of the evolution of the eye is definitely not simple, because the biochemical basis of even the simplest light sensitivity is highly complex. For this reason, claims Behe, typical scenarios of eye evolution do not really solve the difficulties involved, since the small steps described in these scenarios are giant leaps on the molecular level. Because of the great complexity of biochemical systems, Behe claims that evolution must have proceeded through non-functional intermediate steps, and can only be plausible if it was guided by intelligent design. His argument has emerged as one of the most commonly discussed arguments of the ID movement (Kojonen 2016, 56–67).

According to Behe (2006, 39), an "irreducibly complex" system is "a single system composed of several well-matched, interacting parts that contribute to the basic function, wherein the removal of any one of the parts causes the system to effectively cease functioning."[9] According to

Behe, a system for water propulsion, for example, requires a minimum of three parts: a paddle, a motor, and a part connecting the two. The parts also have to fit together tightly, being "well-matched." He presents many different examples of systems, and evolutionary biologists have responded to Behe on each count, arguing for the feasibility of evolution (Kojonen 2016, 63–64).

Here I will comment on the discussion by focusing on the example of the bacterial flagellum, which quickly emerged as the most discussed example of such a system, being like an outboard motor which propels bacteria forward (while also having other functionality). However, for the sake of fairness, I should note that Behe himself saw his argument as cumulative, with the multitude of examples buttressing the case for the improbability of evolution. For instance, he claims that the flagellum itself only functions as part of a network of interlinked irreducibly complex machines, already requiring other irreducibly complex machines for its construction (Behe 2007, chapter 5).

As I argued in the *Intelligent Design Debate and the Temptation of Scientism* (Kojonen 2016, 56–67), the mainstream scientific community has rejected the idea that more than standard evolutionary processes are required to explain the bacterial flagellum or the other examples of complex "machines" described by Behe. Moreover, Behe's definition of irreducible complexity clearly needs modification (Dembski 2004, 2002a, 285; Draper 2002; Boudry et al. 2010). Many different versions of the bacterial flagellum exist in nature, and thus it is difficult to argue that all of its parts are necessary for its current function. Behe (2006, 39, 72–73) does admit that all of the parts of his systems are not necessary, nor are the functions of all parts even known. Hence, irreducible complexity has to refer merely to some core parts and the function of that specific system, with additional parts making the system more robust.

However, the more important critique focuses on the possibility of evolving even systems whose current primary function requires a certain minimum number of parts. Even if the current function requires a certain number of minimum parts, this still does not mean that the system and its parts could not have previously had other functions. Behe argues that irreducible complexity is a barrier to "direct evolutionary pathways," by which he means evolution that improves the system's existing core function step by step, as in the idea that eyes evolve with slight mutations that improve vision cumulatively. Here, the irreducible complexity of a certain function could pose a problem. Since the core function of an irreducible complexity emerges only after all necessary parts are in place, it cannot plausibly be evolved in this direct way. After all, natural selection cannot select for a function that emerges only after all of the parts are in place, because selection cannot look to the future. Instead, the gradual evolution of the parts of a system like the flagellum would have to be favored by natural selection for some other reason, not because of increases in mobility.

Most of Behe's critics have agreed with this point, but have argued that this does not at all prevent one primary way that evolution is thought to happen: the modification of existing systems to fulfill some new function, *exaptation* (e.g., Miller 2002, 132–136; Orr 1996; Gishlick 2004). Behe calls this indirect evolution. Even without contributing to the system's current primary function, the precursors of that system could have had other functions which could have been favored by selection. The evolutionary history of the bacterial flagellum, for example, could be very complex, with similar parts serving in many slightly different systems with different functions over the course of evolution. Such intermediate steps would then enable the emergence of the more complex function. Though a complete evolutionary explanation for the bacterial flagellum is still missing, critics of Behe have argued that approximately 90% of the parts of the flagellum are similar (or homologous) to parts that have other uses, and this gives grounds for constructing a plausible evolutionary explanation for its evolution. The type III secretion system, for example, has been argued to represent a viable precursor system to the flagellum (Musgrave 2004; Pallen and Matzke 2006).

As I have noted (Kojonen 2016, 64–67), though not many critics have engaged with this part of Behe's argument, Behe (2006, 40, 66) himself considers the co-option answer in *Darwin's Black Box*:

> Even if a system is irreducibly complex (and thus cannot have been evolved directly), however, one cannot definitively rule out the possibility of an indirect, circuitous route. As the complexity of an interacting system increases, though, the likelihood of such an indirect route drops precipitously. And as the number of unexplained, irreducibly complex biological systems increases, our confidence that Darwin's criterion of failure has been met skyrockets toward the maximum that science allows.

Here (and elsewhere, e.g., Behe 2001) Behe admits that an irreducibly complex system could in principle evolve in the tinkering, indirect fashion that Behe's critics point to. However, he claims that, as the complexity of the system increases, the probability of such evolutionary accounts decreases. Because the proteins must fit together, the parts must be modified before serving in the new function. Thus, "analogous parts playing other roles in other systems cannot relieve the irreducible complexity of the new system; the focus simply shifts from 'making' the components to 'modifying' them" (Behe 2006, 112–113). Orr (1996), who is otherwise critical of Behe's work, surprisingly agrees with this criticism: "we might think that some of the parts of an irreducibly complex system evolved step by step for some other purpose and were then recruited wholesale to a new function. But this is also unlikely. You may as well hope that half your car's transmission will suddenly help out in the airbag department. Such things might happen very, very rarely, but they surely do not offer a general solution to irreducible complexity." Here, the appeal to our common human experience of designing things supports the inference that creating complex teleological order is difficult. There is, indeed, quite a bit of serendipity in parts useful for one purpose being so easily adaptable to another role.

Draper (2002) homes in on the crucial question: Are the requirements for each individual part really as strict as Behe claims? If biological parts are more malleable than Behe assumes, so that less specificity is required for fulfilling their roles, then Behe's argument against co-option fails. Debunking Behe's argument, then, depends on the details of how proteins work and how difficult it is to transition from one form to another, somewhat similar, form. Then, a continuous series of functional forms, leading from no flagellum to a flagellum, must exist so that no change is too large for natural selection to cross, and all modifications can be made. As with Dembski's argument, it does seem plausible that evolving such complex systems is difficult, and the existence of such an evolutionary pathway has stringent conditions. But difficult or not, it is possible that nature does allow it. Behe thinks that the existence of such pathways is unlikely, but the existence of such pathways is fundamentally an empirical question. The existence of similar parts in other systems, for example, does provide supporting evidence for evolvability (Musgrave 2004; Pallen and Matzke 2006).

Suppose for the sake of the argument that Behe is partially correct: complex machinery exists in nature and is difficult to evolve. Nevertheless, suppose that his critics are also correct, and the evolution of such complexity through Darwinian mechanisms actually happened. Given these premises, a theistic evolutionist could well argue that the irreducible complexity argument merely shows how demanding the conditions for evolvability are, and how much fine-tuning evolution actually requires. In a universe designed to allow for evolution, such serendipity could be expected, rather than being unlikely. Hence, Behe's argument could simply reveal the extent to which fine-tuning is required by evolution. Indeed, the ability to adapt previously evolved features to a new function is commonly discussed in the literature under the term *pre-adaptation*—a term that Gould and Vrba (1982) wanted to replace with the previously mentioned *exaptation*, since talk of pre-adaptations seemed too suggestive of teleology. The plausibility of seeing the evolutionary process as somehow directional or constrained is also suggested by many other factors, such as the evidence from protein evolution, to which I now turn.

Protein Evolution

How rare are functional protein forms compared to the totality of possible forms? Through estimates of just how rare functional proteins are, all agree that the numbers of possible arrangements of amino acids are hyper-astronomical. As Wagner (2016) points out, a library containing "every possible amino acid string that is 500 letters long would contain more than 10^{600} texts—a one with 600 trailing zeros. That vastly outnumbers the atoms in the visible universe." The sequence of amino acids determines the three-dimensional form of the protein. The stringency of the requirements for protein tasks, particularly for their active sites and for fitting in with other proteins in larger wholes, is generally thought to give grounds for supposing that functional proteins will be rare in comparison to non-functional ones (Hecht et al. 2004). However, numerical estimates vary greatly. On the more pessimistic side, ID proponents and biochemists Axe and Gauger argue (Axe 2004; Gauger and Axe 2011) that functional protein folds are very rare, so that based on their study of the sensitivity of the beta lactamase enzyme to mutation, only about 1 out of 10^{64} protein folds are functional. However, others argue that functional proteins are much more common. A similar study (Taylor et al. 2001) done on the chorismate mutase enzyme yields a figure of 1 out of 10^{23} of functional to non-functional sequences—still exceedingly rare, but much more common than in the case of the protein studied by Axe. It seems, then, that the specificity required for achieving a functional amino acid sequence varies somewhat for different proteins (Tian and Best 2017).

Functionality is much harder to achieve than stability. If we do look at just the requirement of achieving a stable three-dimensional form, without requiring any functionality, then traditional estimates indicate that this can be done by perhaps one in ten thousand amino acid sequences (Ferrera and Wagner 2011, 1).[10] However, the overall trend of research in recent years seems to point to the conclusion that the requirements of stable proteins are also much more stringent than previously assumed. Tokuriki and Tawfik (2009) claim that in recent years, "the accepted paradigm that proteins can tolerate nearly any amino acid substitution has been replaced by the view that the deleterious effects of mutations, and especially their tendency to undermine the thermodynamic and kinetic stability of proteins, is a major constraint on protein evolvability—the ability of proteins to acquire changes in sequence and function."

The extreme rarity of functional forms has recently emerged as a new primary argument of the ID movement. This is a natural consequence of the way the debate has developed: as I argued previously, Dembski's case against evolution depends on irreducible complexity, and irreducible complexity depends on the difficulty of protein evolution (as well as the difficulty of modifying proteins and constructing complex organs from them). Meyer (2013, 199), for example, argues that "since natural selection does nothing to help *generate* new folded functional sequences, but rather can only *preserve* such sequences once they have arisen, random mutation alone must search for the *exceedingly* rare folded and functional sequences within the vast sea of combinatorial possibilities." Meyer's argument is that due to the extreme rarity of functional sequences, new protein folds are too difficult for random mutations to find in the finite time that is available for evolution.

To defend the plausibility of evolution, we could take the line that functional forms are actually more common than Meyer assumes. And since estimates based on empirical studies vary, it does seem credible to assume that not all proteins require as much specificity as Axe (2004) claims. However, whatever the exact numbers end up being, most estimates still admit to the extreme rarity of functional forms. The more crucial point that needs to be made in response is that what matters is not just the rarity of functional forms, but also their closeness in the "biological hyperspace" of functional forms, meaning the "distance" in mutations that is required to traverse between these forms.[11] If functional forms are close to one another, then producing a new protein form or function would not require evolution to search through the entire vast realm of all possible arrangements of amino acids. Rather, evolution would only need to search through the adjacent space of possible forms to find viable new forms. This is a far easier task, and if functional forms are arranged in such a way, then this would explain how evolution is possible despite the rarity of functional forms.

Separately from the discussion over ID and irreducible complexity, Wagner (2014) argues that this closeness of functional forms can be demonstrated both empirically and through simulations. Wagner uses the metaphor of a "library of forms" to illustrate the relations of possible biological forms, and how evolution moves between them. It is as if different possible protein forms are like texts in a library, and one must travel in this library always from one text to an adjacent text. The library contains all possible combinations of amino acids, each represented by a single text. According to Wagner, it could in principle be that the meaningful texts—the viable protein forms—in such a library were like "stars in our universe, islands separated by vast expanses of dark empty space" (Wagner 2014, 100). In that case we would be in the situation described by the aforementioned ID proponents, and evolution by natural selection would be impossible.

Through empirical and computational work on proteins, Wagner argues, also supported by the work of his research team (Aguilar-Rodriguez and Wagner 2009; Aguilar-Rodriguez et al. 2017, 2018; Fortuna et al. 2017), that different proteins are in fact closely connected, in a gigantic "genotype network" of functional forms that evolution can traverse. This means that evolution only needs to explore a minute portion of the space of possible forms in order to find function. This, according to Wagner (2014, 186), is a crucial part of explaining "how blind search produces life's immense diversity." He elaborates that the usefulness of such a network for evolution depends on two additional conditions: diversity and robustness. Diversity of function in nearby forms is essential, because if the nearby forms all did the same thing, then they would not provide any new possibilities for evolution. In effect, the "library" must have volumes on different topics mixed together. Robustness—the ability of protein forms to tolerate changes—is also important, because it allows for exploratory, neutral changes in the current forms without removal of the current function. According to Wagner, both conditions are met by the arrangement of life's metabolic library. Thus, the library of forms is "perfect, uncannily well set up for innovation. It guarantees eternal learning and innovability" (Wagner 2014, 106). Many others have joined in arguing that features of the space of possible forms can influence, and even determine, evolution. For example, Ahnert (2017) argues that "Characteristics such as the skewed distribution of neutral network sizes and their shape in genotype space can strongly determine evolutionary outcomes. [...] More generally speaking this result implies that the accessibility of phenotypes may be just as important a determinant of evolutionary outcomes as the pressures of natural selection."

According to this view, then, the possibility of evolution depends on the features of the space of possible forms, where all the forms must be arranged in a way that makes an evolutionary search through it possible. This argument shows how the preconditions for the working of the "blind watchmaker" of natural selection can indeed be satisfied by nature in the case of protein evolution, despite an extreme rarity of functional forms. Behe (2019, 112) argues that Wagner does not yet solve the puzzle of evolving irreducible complexity, arguing that "it doesn't even try to account for the cellular machinery that is catalysing the chemical reactions to make the needed components."[12] However, suppose that, in the case of the bacterial flagellum, though the vast majority of possible arrangements of biological proteins are non-functional, there nevertheless exists a series of possible functional forms, little "machines" that happen to contain increasing numbers of the flagellum's vital parts while still serving some other function. This then would allow for the seamless transition from no flagellum to a flagellum over time, through small successive steps. In this manner, by moving through such a suitable library of forms, the blind process of evolution would have the ability to produce even the most complex structures without the intervention of a designer. This is the kind of fine-tuning of the landscape of forms that seems to be required to evolve the kind of biological order described by Behe.

It seems, then, that defending the power of the evolutionary mechanism requires assuming that the landscape of possible biological forms has some fairly serendipitous properties. This, then, arguably moves the problem of design back into the laws of nature by showing that biological evolution is just as dependent on the fulfillment of very precise preconditions as the previously discussed genetic algorithms. The ability of evolution to generate teleology appears to depend on teleology, just as the Thomists suspected. Hence, the discussion on biological design arguments itself provides some reasons to think that evolution must be fine-tuned in order to work, and even anti-evolutionist arguments show ways in which evolution could have failed to work. The dependency of evolution on such wider teleology is further supported by many additional lines of evidence, to which I will now turn.

4.3 Teleology All the Way Down?

Laws of Form and Convergence

Based on the preceding, it seems readily apparent that natural selection and mutation will not work in just any environment. The addition of such preconditions as the "library of forms," itself an emergent consequence of the laws of chemistry and physics, seems to shift much of the explanatory work away from natural selection and mutation, into the environment which makes these possible. Here, chance mutations and natural selection act more in the role of a "search engine," looking through this space of possible forms. As biologists increasingly talk of "laws of form" underlying evolutionary development, the role of natural selection and mutation in explaining biological form seems comparatively less all-encompassing. The resulting picture comes quite far from the old view that all one needs for evolution is differential survival, reproduction, and heritability, evolution then following from these as almost a logical necessity. Again, the basic trend of research has been visible for some time. Sean Carroll (2001, 62) wrote that:

> Life's contingent history could be viewed as an argument against any direction or pattern in the course of evolution or the shape of life. But it is obvious that larger and more complex life-forms have evolved from simple unicellular ancestors and that various innovations were necessary for the evolution of new means of living. This raises the possibility that there are trends within evolutionary history that might reflect the existence of general principles governing the evolution of increasingly larger and more complex forms.[13]

Alexander (2018, 93) remarks that such continuing developments seem almost like a return to the views of the pre-Darwinian biologist Richard Owen (1804–1892), who famously argued that biological forms are mere images of archetypal forms in the mind of God: "by the late twentieth century, ideas that were looking remarkably like the *Bauplan*, albeit shorn of its original metaphysical baggage, were once again becoming common within mainstream biology." According to Alexander (2018, 103), "we cannot avoid the constrained features of the evolutionary process, dependent ultimately upon the laws of chemistry and physics. Constraint and convergence are all around us. Nomic (lawlike) regularity is a consistent feature."[14] McLeish (2020, 56) similarly argues that "the size, shape and topology of the genotype-phenotype map, together with the random microscopic mutation, conspire, as do the boundary conditions for fluids, to generate a determined evolutionary history with a high degree of certainty."

Gould (1989) famously argued that the course of evolution is highly contingent: if we were to "rewind the tape of life" and play it again from the beginning, the result would most likely be very different. Paleontological discoveries, particularly in the Cambrian era, had revealed a large variety of different animal types, and it seemed likely to Gould that the survival of the lineages that led to the present types of species was more due to evolutionary accidents than any historical necessity. Even otherwise successful species can become extinct due to sudden changes in the environment, or disasters like the meteorite impact apparently responsible for the demise of the dinosaurs 65 million years ago. In another famous metaphor, Gould (1996) compared the evolution of a species to the random walk of a drunkard, leaning against the "wall" of minimal complexity required for survival, but otherwise walking randomly. Even though not all would emphasize the contingency of evolution to the degree that Gould did, it has long been a mainstream interpretation of evolution that despite the tendency of organisms to adapt to their local environments, there is nevertheless no overall direction to the process, other than perhaps a tendency to diversify (Ruse 2013).

However, while the debate over the level of contingency in the evolutionary process is ongoing, the previously mentioned data from protein evolution already seems to show that there are some constraints to what kind of results the process can produce, and thus some directionality. As Weinrich and colleagues (2006, 111) conclude: it seems that Darwinian evolution can follow only very few mutational paths to fitter proteins, and thus "the protein tape of life may be largely reproducible and even predictable." This is also supported by the study of situations in which the features of the fitness landscape seem to prevent evolution. In the case of RNA evolution, Jiménez and colleagues (2013) argue that "the landscape is composed of largely disconnected islands of active sequences. Natural selection under these conditions would be constrained to local exploration of sequence space." This same result can also be supported by studies of evolutionary constraint (Koonin and Wolf 2010) and situations in which evolution is hindered by factors like this (Pressman et al. 2019).

This seems to lead to the view that the course of evolution is indeed constrained in many respects by the laws of physics and chemistry. ID-friendly Denton and colleagues (Denton and Marshall 2001; Denton et al. 2002) argue, similar to Wagner, that the biochemical data even support what they term a "platonistic" view of protein forms. They claim that the basic possible folds are a direct consequence of the laws of physics, and that these act as stable nodes toward which evolution gravitates. Thus "Protein folds found in nature represent a finite set of built-in, platonic forms. Protein functions are secondary adaptations of this set of primary, immutable, natural forms" (Denton et al. 2002, 334).

Denton (2013, 2016) explicitly adopts a "structuralist" view of evolution, meaning the idea that biological form is the consequence of the laws of physics and chemistry, merely discovered by evolution. Structuralism is a minority position among biologists, but has found many champions during the twentieth century, such as Brian Goodwin, who famously argued that we might eventually be able to formulate something akin to a periodic table of life, in which laws of form are an indispensable explanatory component of evolution, in addition to other factors like mutation and selection. Goodwin's (2001, 114) dream was that "Biology would begin to look a little more like physics in having a theory of *organisms* as dynamically robust entities that are natural kinds, not simply historical accidents that survived for a time."

In addition to the evidence from biochemistry cited by Denton, Wagner, and others, support for an understanding of evolution in which "laws of form" play a significant role also comes from convergence, notably discussed by Simon Conway Morris (2005, 2009, 2010, 2015) and George McGhee (2011). Convergence refers to the independent evolution of the same biological outcome in two or more different lineages, beginning from different starting points. For example, dolphins and sharks have similar streamlined bodies and dorsal fins, even though dolphins are mammals and sharks are fish. Paddle-shaped limbs for swimming have evolved independently seven times, and a structure as complex as the eye has evolved independently 49 times (although some regulatory genetics are shared). Examples of convergence are ubiquitous in biology. Willmer (2003, 33) observes that such "convergent evolution is prevalent at all levels of organismal design—from cell chemistry and microstructure to cell types, organ systems, and whole body plans."[15]

For Conway Morris, the implications of convergence are clear: evolution has a direction, and inevitably tends toward certain types of forms: "at whatever level of biology one considers there will be loci of persistent biological stability that will act as irresistible attractors" (Conway Morris 2015, 33). He admits that contingent events like natural catastrophes can have a major effect on evolution. There are also more possibilities than evolution has in practice found. But these are overshadowed by the necessity driving the process. The roads of evolution are, just as in the previously mentioned example of protein evolution, "exceedingly narrow," and "we need constantly to remind ourselves that practically all the alternatives are not worlds that might have been, but worlds that will never be" (Conway Morris 2015, 221). "Evolution perforce must follow a metaphorical map of highly restricted possibilities," (Ibid., 24) and is "simply the search engine that allows the universe to become self-aware" (Ibid., 295). Thus understood, evolution searches through the space of possible forms, *discovering* biological forms, rather than *creating* these forms out of nothing. Conway Morris also argues that the previously mentioned phenomenon of pre-adaptation, the fortuitous usefulness of previously evolved features for further evolution, fits well with a teleological understanding of evolution. Indeed, insofar as the aforementioned points about the fine-tuning of the library of forms are correct, then they could help explain the phenomenon of convergence pointed out by Conway Morris. This possibility is discussed by Louis (2016)—though he emphasizes that we have not yet determined the extent to which the structure of the genotype-phenotype map explains evolutionary trends. However, Conway Morris presents grounds for arguing in the other direction, for seeing convergence as evidence that there indeed are laws of form which direct evolution.

The evidence that there are "laws of form" that have some importance for explaining biology appears strong. However, as noted, there is still much debate to be had over just what the extent of this is, and not all take the course of evolution to be as strongly directional as Conway Morris and others do. For example, McGhee (2011, 271) agrees that "the view that evolution is entirely historically contingent, and thus unpredictable (and nonrepeating), is demonstrably false." He also finds the idea of a "biological possibility space" or "morphospace" useful, and argues that the features of this space are open to analysis using the tools of theoretical biology (McGhee 1999). However, while he thus agrees with the general outlines of the picture given by Conway Morris, he considers that uncertainty is created by the phenomenon of evolutionary constraint: the dependency of future evolutionary paths on earlier evolutionary events. In analyzing theoretical morphospace, it becomes clear that there are some possible biological forms which are nevertheless in practice impossible to realize for life on Earth. For example, the form of a dragon—a reptile with functional fore and hind limbs as well as wings—is theoretically possible, but has not been realized by Earth life, most probably because of the difficulty of evolving a further pair of limbs and then transforming these into wings. The evolutionary pathway to such a creature is most likely simply missing (McGhee 2011, 19–20).

To explain McGhee's critique through Wagner's metaphor of the library of forms: Earth life must start from a certain point in the library, and can discover only those functional forms that are connected, via functional intermediates, to this original starting point. However, sometimes there are many paths open to the organisms, and once evolution has headed on a certain path, it becomes difficult to evolve in the other directions. The form that biological organisms evolve constrains the possibilities of further evolution. Thus, there could in principle be other functional forms that are situated outside the network of forms that are accessible for current Earth life. Conway Morris and McGhee see convergence at work already in the Cambrian forms of life, which Gould thought demonstrated that the variety of current life hinges on contingent events far back in evolutionary history. However, as McGhee points out, as long as the origin of life remains unexplained, it is possible that there are different starting points for life, which would lead to different forms. Thus, convergence only allows us to state that the evolution of Earth-like life is heavily influenced by functional constraints.

The origin of life cannot be discussed here in depth. The hypothesis of the origin of life as a pure chance event, as opposed to a combination of chance and necessity, is rejected as unexplanatory by the majority of scientists (Bertka ed. 2009; Fry 2000). Thus, it seems in principle possible that the origin of life will ultimately prove to be highly non-random, either as an event determined by as yet unknown, fine-tuned laws of nature or as a creative miracle. Though current theories are still largely incomplete, the immense improbability of a chance origin, the relatively quick origination of life on Earth after its cooling, and the relative optimality of the genetic code seem to point away from chance as the explanation for the origin of life (Wichmann and Ardern 2019). Melissa Ilardo and colleagues (2015) similarly argue that "the amino acids used for constructing coded proteins may represent a largely global optimum, such that any aqueous biochemistry would use a very similar set." Discussion of other potential chemical bases for life is ongoing within astrobiology. Some still continue to argue for a chance origin of life. Koonin (2007) even goes so far as to use the multiverse hypothesis to justify belief in a chance origin: if there is an infinite multiverse, then perhaps even a hyper-astronomically unlikely origin of life is bound to happen in some universes just by chance.

Therefore, it seems likely that consideration of the origin of life will tend to support, rather than undermine, the account of Conway Morris and others. However, for the sake of argument, suppose that there are at least several very different, still viable forms of life utilizing a different chemical basis from Earth life. Perhaps silicon-based life, for instance, will turn out to be viable. In that case, '"the laws of biology' could be different for those life forms, and they might evolve forms that are totally unlike those seen on Earth" (McGhee 2011, 276). Ultimately, figuring out the origin of life and studying possible non-Earth life would help test the conclusions we draw from convergence based on Earth life. Thus, for McGhee, our knowledge of the evolutionary process is still too incomplete to make great overall judgments. While there is evidence of directionality in the evolution of Earth-based life, he concludes that "it has not been proved that biological evolution is inevitable, that it has a destination" (McGhee 2011, 274–275).

Determinism and Indeterminism

Whereas McGhee's critical analysis of convergence still mostly aligns with Conway Morris' conclusions, and in my view he argues strongly against accounts emphasizing the limitlessness of the possibilities available to evolution (McGhee 2011, 266–269), others continue to side more with Gould's emphasis of contingency (Conway Morris and Gould 1998), although many of Gould's specific empirical arguments have been overturned by the research in the decades since (Erwin 2016). Kauffman (2008, 130), for example, argues that "the evolution of the biosphere is radically nonpredictable and ceaselessly creative," and that because the difficulty of attaining all possible states "increases as the complexity of 'objects' (molecules, protocells, cells, organisms) increases, I strongly doubt Conway Morris' interesting 'saturation' claim. This remains an important open question" (Kauffman 2013, 168). By saturation Kauffman means the idea that natural selection has produced most of the phenotypes that can be evolved. If there are always many more possibilities open to evolution than are actualized by evolution, then evolution would actually be unpredictable, even if its possibility depends on the kind of library of forms that Wagner describes. However, Kauffman also believes that many features of life are the direct result of self-organizational principles working at all levels of nature, rather than the result of natural selection. Because of this, for Kauffman (1996), life is fundamentally "at home" in the cosmos, rather than being just an accidental by-product. This kind of directionality in nature is consistent with a teleological understanding of nature, and could plausibly be combined with the arguments of structuralists like Denton. As I will argue further below, it does not seem that anything in the dependency of evolution on laws of form requires the claim of saturation that Kauffman criticizes.

Related to this, philosopher of biology Massimo Pigliucci argues that Wagner's idea of a platonic library of forms as predating evolution multiplies ontologies unnecessarily, and it is hard to understand in what sense the forms could exist before being produced by evolution. For Pigliucci (2016), "the library doesn't exist before natural selection 'walks' through it. The analogy is misleading: It is better to think of a library that is created (and partially destroyed) moment by moment as life evolves." Pigliucci's modification of the metaphor captures the insight that the usefulness of different forms is also dependent on the conditions of each ecosystem, and thus it can well be argued that the routes available to evolution are also continually changing. Nevertheless, he does not provide a response to the arguments underlying Wagner's library of forms, or the dependency of different possible forms on the laws of physics. While fitness landscapes will change to some degree, the stability and physical functionality of certain arrangements of matter depend on unchanging physics. Elsewhere, Pigliucci himself (2012) has also argued for the usefulness of the concept of theoretical morphospace, which McGhee uses to argue for the constrained nature of the evolutionary process.

The traditional critique of structuralism has been that, as Griffiths (1996) puts it, "nothing in the phenomena suggests the sort of manageable periodic table of organismic forms that would be necessary to make the structural explanations of form […] genuinely explanatory. The generic forms that exist in nature may be a tiny subset of the possible generic forms that could have been created by the historical design of alternative developmental systems." Structuralists tend to argue that viable forms are vastly outnumbered by non-viable forms. In contrast,

Griffiths argues that the diversity of biological form itself, particularly in the Cambrian explosion, suggests that current biological life indeed only inhabits a tiny portion of the space of viable forms, and that because of the abundance of different viable forms, historical Darwinian reasons for the existence of the forms that obtain should be preferred.

In his critique of Goodwin, Dennett (1995b) presses this point by extending Goodwin's argument to the engineering world: "I don't think there's anybody who thinks there are deep, fundamental laws of automotive engineering, but when Goodwin says that there are laws of form in biology, he's making what I view as an equally implausible claim." Dennett's position, in defense of the historical Darwinian view, is that the laws of nature are indeed necessary for the possibility of evolution, but they simply give the basic rules that then allow for the creation of an endless variety of forms, just as the laws of physics allow for the creation of an endless number of human artifacts. While the laws of nature make certain arrangements work better, much of the specifics are left to historical processes. Just as there are different ways to build a vehicle, so there are also many different ways to build a successful biological organism. Moreover, evolution is almost endlessly creative, and can repurpose and modify even similar structures in an enormous variety of ways. Witness, for example, the multitude of uses of the same basic limb structure in vertebrates. Why should we think, then, that the laws of form have a central explanatory role in explaining biology over against contingent and historical factors?

However, Dennett's analogy is not fully convincing, since evolution is much more restricted than human engineering. Unlike human engineers, who can design machines like aircrafts using top-down and forward-looking thinking, with no requirement that each stage in the construction must already be fully functional for some purpose, evolution must create its products by stepwise tinkering where intermediate states must not hurt chances of survival. This difference grounds the necessity of the library of forms as a precondition of evolution. However, leaving that aside, the critique does allow us to construct an alternative perspective to structuralism. Supposing that the landscape of forms is not as fine-tuned and as restricted as the structuralists and others like Wagner claim, then the role of contingent historical factors in explaining biological order becomes more prominent than if the constraints given by the laws of form are more pressing.

Most evolutionary biologists today seem to take the view that there is evidence for both constraint and contingency in the evolutionary process.

The most extreme claims of contingency have not held the test of time (McGhee 2011; Erwin 2016), but there is also evidence of the importance of historical factors in shaping the direction of evolution, and room for further research on the repeatability of evolution (Kacar 2016). Thus, it seems to me that those emphasizing the existence of laws of form have presented strong evidence; the issue is simply the relative extent to which these influence the course of evolution. Here I can distinguish two separate questions at stake in the critique: (1) whether the possibility of evolution is tightly dependent on possible "laws of form" and the suitability of the space of possible forms for evolution, and (2) whether evolution is highly predictable, with no room for historical contingency. Consider again Wagner's analogy of the library of forms. In this metaphor, evolution is highly dependent on the existence and arrangement of these possible forms that result from the basic laws of chemistry and physics. The features of the space of possible forms are a central factor that needs to be understood to explain the possibility of evolution. Nevertheless, nothing about the metaphor requires that the search engine of evolution has discovered anything more than a minute fraction of all forms that exist in the library, or that historical contingency could not play a large role in explaining which of the myriad possible forms end up being discovered. Even supposing that life could have developed in a widely different direction, the possibility of evolving life still depends on the space of possible forms being structured in a suitable way.

The debate over convergence and the predictability of evolution does have implications for the wider theological discussion over whether evolution has a God-given direction, and if we can have evidence of this (Alexander 2018; Glass and McCartney 2014, 355). In the philosophical discussion over whether evolution can be spoken of as a "progression," attention is usually given to the observed increase of complexity, variety, and brain size over evolutionary time. These are already significant from a theological point of view. As Ward (2004, 269) argues, "It is undeniable that organisms have developed from unicellular organisms to beings with central nervous systems and brains. If a necessary condition of the actual existence of values is the existence of consciousness, understanding and feeling, then there is here a development of values from a cosmos in which there were no actual values (except, perhaps, those experienced by God in contemplating the beauty of nature). That is a strong enough sense of direction, development and progress for any theist." The considerations analyzed in this chapter contribute to this argument, and seem to support expanding the fine-tuning argument (Ardern 2018). However, the argument from fine-tuning and directionality is different from the one under consideration in this book, since the biological design argument, as understood here, is not based on the sense of progression in evolution. Rather than focusing on the overall purpose and directionality in evolution, the argument is based on the immanent apparent teleology of biological organisms themselves as the basis of a perception, sign, and/or evidence of the Creator.

Thus, the uncertainties regarding convergence and the origin of life may affect the relative role of laws of form and the extent to which evolution is predictable, and they may affect estimates about overall purpose in evolution. Nevertheless, the general dependence of evolution on preconditions can be affirmed by both the structuralists and their critics, and it is this aspect that helps in my salvaging the biological design argument within a theistic evolutionist understanding.

4.4 Conclusions

In popular-level discussions, the mechanism of natural selection, working on hereditary random mutations, is often argued to almost inevitably result in the increase of complex order. If we have inheritance, variation, and differential survival, then evolution will happen. Nevertheless, both simulations and empirical studies of evolution, as well as the defenses of evolutionary explanations against critique, show that other factors are also necessary. It is easy to construct simulations where evolution cannot proceed beyond a certain point, for instance by making the steps required for evolution too great. This is apparent both in the responses to anti-evolutionist arguments and in other work. In order for evolution to be possible, viable forms must be close enough to each other in the space of possible forms, and must form a network that can be navigated by evolutionary search. The study of protein evolution and convergence, among other factors, provides evidence that the course of evolution is directed by the structure of the space of forms, as well as laws of form, arising as a consequence of the laws of physics. Also, it may be that evolutionary biology could be understood as even hospitable to teleological interpretation, once all elements of the extended synthesis are considered fully. As Hoggard Creegan (2013, 121) argues, "While teleology strained against the grammar of natural selection alone, it is not at all a foreign or difficult concept in the light of convergence, *evo devo*, and epigenetics."

The general point that evolution depends on the wider teleology of the cosmos does not depend on this more detailed case for fine-tuning and laws of form underlying evolution. After all, the dependency of evolutionary processes on the wider teleology of the cosmos was already noted before the discovery of any fine-tuning, for it is clear that evolution depends on the existence of an orderly cosmos. Nevertheless, all of this still adds to the case. The more our understanding of evolution has progressed, the more we find ourselves explaining patterns in evolution by reference to general principles, rather than just contingent historical events (which continue to have a great role in all models). Instead of explaining the appearance of purpose in biology by reference to non-teleological factors, the attempt at explaining design by evolution has succeeded in finding new layers of teleology. The further we study, the more the universe seems to be filled with teleology "all the way down!"

All of this adds to the power of the fine-tuning argument, and seems to support the Thomistic view, mentioned in the previous chapter, about the difficulty of explaining teleology without appealing to teleology. Insofar as evolutionary explanations make an intervening designer for biology unnecessary, the weakening of the case for design on the level of biology thus seems to lead to a corresponding increase of fine-tuning. This is already highly significant for the discussion of teleology, but for the purposes of this book, this is not yet sufficient. The intuition behind biological design arguments is that the order of the end result of evolution—the apparent design, the complex teleology of living creatures—itself provides evidence, or allows for the perception of design. Locating the evidence of design solely on the level of fine-tuning, or on the level of the "library of forms," would mean giving up this intuition. In the next chapter, I argue, however, that the dependence of evolution on such general principles also undermines the opposition of evolutionary explanations to biological design arguments, since they push the explanation for biological teleology at least partly into the metaphysical realm. It is possible to see biological organisms as reflecting the designedness of the entire cosmos, and that the "theists on the street" are not wrong to see them as signs of design.

Notes

1. Quoted in Giberson and Artigas (2006, 32–33).
2. However, Fodor and Piattello-Palmarini argue that descriptions of selection as "selecting for" something need to be drastically revised due to their inappropriate reference to intentionality.

3. Darwin (1860) states along these lines in a letter to Lyell: "The very existence of the architect shows the existence of more general laws; but no one in giving credit for a building to the human architect thinks it necessary to refer to the laws by which man has appeared. No astronomer in showing how movements of Planets are due to gravity, thinks it necessary to say that the law of gravity was designed [so] that the planets shd. pursue the courses which they pursue. I cannot believe there is a bit more interference by the Creator in the construction of each species, than in the course of the planets. It is only owing to Paley & Co, as I believe, that this more special interference is thought necessary with living bodies."
4. The discussion of infinite variations and trials as an alternative was debated extensively in the ancient Greek debates on design and chance (Sedley 2007, chapter 5). Dawkins (1986, 67) also uses the example of infinite monkeys.
5. Many would add, following the previously mentioned discussion of neutral evolution, that not all intermediate steps even need to be advantageous. However, at least the intermediate steps cannot be very disadvantageous.
6. This is from Shakespeare's *Hamlet*, Act 3, Scene 2.
7. The example of the tornado in a junkyard originates with Hoyle (1983, 7).
8. For Wolpert's own critique of Dembski, see Wolpert (2002). However, Häggström (2007a, 229–230) argues persuasively that the NFL theorem can in fact also be applied to co-evolutionary scenarios.
9. Ibid., 39.
10. Brandon and Tooze (1999) speculate that even one out of 100 amino acid sequences may achieve a stable fold, but this seems far too low in comparison to the other estimates. Quoted in Denton et al. (2002, 336).
11. For discussion of these theoretical concepts, see particularly Pigliucci (2010) and McGhee (1999).
12. Axe's (2017, 223) response to Wagner is disappointing: "Being familiar with the subject he deals with, I could tell you why I think he didn't succeed, but in effect I would be asking you to trust me over him, which none of us should find satisfactory. Instead, my whole purpose has been to equip you to trust your own design intuition." Axe thinks that the design intuition coupled with commonsense reasoning should lead us to suspect any model of evolution where complexity arises without design. But this is of little help to someone who thinks that the basic reliability of the design intuition is compatible with evolution, as I am arguing in this book.
13. Quoted in Alexander (2018, 62).
14. It could be argued that Owen's approach is quite distant from the Paleyan tradition of teleology, which sees the evidence of design in the adaptive complexity of life. In contrast, the morphologists, whom Owen represents, saw evidence of design in the repeating patterns of life, reflecting a divine

mind. See Bowler (1977). In contrast, Gray saw evolutionary biology as combining the two strands of Paleyan teleology and morphology, and the salvaged biological design argument attempts the same. Gray (1888a [1874]) wrote: "let us recognize Darwin's great service to natural science in bringing back to it Teleology, so that, instead of Morphology versus Teleology, we shall have Morphology wedded to Teleology."

15. Quoted in Moritz (2019, 303).

References

Aguilar-Rodriguez, J.F., and A. Wagner. 2009. Evolutionary Plasticity and Innovations in Complex Metabolic Reaction Networks (Plasticity and Innovations in Metabolic Networks). *PLoS Computational Biology* 5 (12): e1000613.

Aguilar-Rodríguez, J., J.L. Payne, and A. Wagner. 2017. A Thousand Empirical Adaptive Landscapes and Their Navigability. *Nature Ecology & Evolution* 1 (2): 45.

Aguilar-Rodríguez, José, Leto Peel, Massimo Stella, Andreas Wagner, and Joshua L. Payne. 2018. The Architecture of an Empirical Genotype-Phenotype Map. *Evolution* 72 (6): 1242–1260.

Ahnert, S.E. 2017. Structural Properties of Genotype-Phenotype Maps. *Journal of the Royal Society, Interface* 14 (132): 20170275.

Alexander, Denis. 2018. *Is There Purpose in Biology?* Oxford: Lion Hudson.

Ardern, Zachary. 2018. What Can Biology Tell Us About Genesis? *Capturing Christianity*, June 25. Available at https://capturingchristianity.com/what-can-biology-teach-us-about-genesis/.

Ariew, André. 2007. Teleology. In *The Cambridge Companion to the Philosophy of Biology*, ed. David L. Hull, 160–181. Cambridge: Cambridge University Press.

Axe, Douglas. 2004. Estimating the Prevalence of Protein Sequences Adopting Functional Enzyme Folds. *Journal of Molecular Biology* 341 (5): 1295–1315.

———. 2017. *Undeniable: How Biology Confirms Our Intuition That Life Is Designed*. San Francisco, CA: HarperOne.

Ayala, Francisco. 2010. Darwin's Explanation of Design: From Natural Theology to Natural Selection. *Infection, Genetics and Evolution* 10 (6): 840–843.

Behe, Michael J. 2001. Reply to My Critics: A Response to Reviews of *Darwin's Black Box: The Biochemical Challenge to Evolution*. *Biology & Philosophy* 16 (5): 685–709.

———. 2006. *Darwin's Black Box: The Biochemical Challenge to Evolution*. 10th Anniversary ed. New York, NY: The Free Press.

———. 2007. *The Edge of Evolution: The Search for the Limits of Darwinism*. New York, NY: The Free Press.

———. 2019. *Darwin Devolves: The New Science about DNA That Challenges Evolution*. San Francisco, CA: HarperOne.

Berhow, Michael. 2019. *Dysteleology: A Philosophical Assessment of Suboptimal Design in Biology*. Eugene, OR: Wipf and Stock.

Berlinski, David. 2004. A Scientific Scandal. *Commentary*, April. Available at https://www.commentarymagazine.com/articles/david-berlinski/a-scientific-scandal/.

Bertka, Constance M., ed. 2009. *Exploring the Origin, Extent, and Future of Life: Philosophical, Ethical, and Theological Perspectives*. Cambridge: Cambridge University Press.

Blancke, Stefaan, Maarten Boudry, and Johan Braeckman. 2011. Simulation of Biological Evolution Under Attack, But Not Really: A Response to Meester. *Biology & Philosophy* 26 (1): 113–118.

Boudry, Maarten, Stefaan Blancke, and Johan Braeckman. 2010. Irreducible Incoherence and Intelligent Design: A Look into the Conceptual Toolbox of a Pseudoscience. *The Quarterly Review of Biology* 85 (4): 473–482.

Bowler, Peter J. 1977. Darwinism and the Argument from Design: Suggestions for a Re-evaluation. *Journal of the History of Biology* 10 (1): 29–43.

Bowler, Peter. 2009. *Evolution: The History of an Idea*. 25th Anniversary ed. Berkeley, CA: University of California Press.

Brandon, C., and J. Tooze. 1999. *Introduction to Protein Structure*. 2nd ed. New York, NY: Garland Publishing Inc.

Browne, Janet. 2010. Asa Gray and Charles Darwin: Corresponding Naturalists. *Harvard Papers in Botany* 15 (2): 209–220.

Carroll, Sean B. 2001. Chance and Necessity: The Evolution of Morphological Complexity and Diversity. *Nature* 409: 1102–1109.

Conway Morris, Simon. 2005. *Life's Solution: Inevitable Humans in a Lonely Universe*. Cambridge: Cambridge University.

———. 2009. The Predictability of Evolution: Glimpses into a Post-Darwinian World. *Naturwissenschaften* 96: 1313–1337.

———. 2010. Evolution: Like Any Other Science It Is Predictable. *Philosophical Transactions of the Royal Society B* 365 (1537): 133–145.

———. 2015. *The Runes of Evolution: How the Universe Became Self-Aware*. West Conshohocken, PA: Templeton Press.

Coyne, Jerry. 2007. The Great Mutator. *The New Republic* 39: 38–44.

Darwin, Charles. 1860. To Charles Lyell, 17 June. *The Darwin Correspondence Project*. Available at https://www.darwinproject.ac.uk/letter/?docId=letters/DCP-LETT-2833.xml.

———. 1861. To Charles Lyell, 1 August 1861. Darwin Correspondence Project, Letter No. 3223. Available at https://www.darwinproject.ac.uk/letter/DCP-LETT-3223.xml.

———. 1868. *The Variation of Animals and Plants Under Domestication*. Vol. 2 vols. London: John Murray.
———. 1958 [1887]. In *The Autobiography of Charles Darwin*, ed. Nora Barlow. London: Collins.
———. 2008 [1859]. *On the Origin of Species by Means of Natural Selection, Or, the Preservation of Favoured Races in the Struggle for Life. 2nd Edition (1859)*. Oxford World's Classics. Oxford: Oxford University Press.
Dawes, Gregory. 2007. What *Is* Wrong with Intelligent Design? *International Journal for Philosophy of Religion* 61 (2): 69–81.
Dawkins, Richard. 1986. *The Blind Watchmaker*. Harlow: Longman.
Dembski, William A. 1998. *The Design Inference: Eliminating Chance Through Small Probabilities*. Cambridge Studies in Probability, Induction and Decision Theory. Cambridge: Cambridge University Press.
———. 2002a. *No Free Lunch: Why Specified Complexity Cannot Be Purchased Without Intelligence*. Lanham: Rowman & Littlefield Publishers.
———. 2002b. Obsessively Criticized but Scarcely Refuted: A Response to Richard Wein. Design Inference.com. Available at: https://billdembski.com/documents/05.02.resp_to_wein.htm.
———. 2004. Irreducible Complexity Revisited. Unpublished Paper. Available at https://billdembski.com/documents/2004.01.Irred_Compl_Revisited.pdf.
Dembski, William A., and Robert J. Marks II. 2009a. Conservation of Information in Search: Measuring the Cost of Success. *IEEE Transactions on Systems, Man and Cybernetics A, Systems & Humans* 5 (5): 1051–1061.
———. 2009b. Life's Conservation Law: Why Darwinian Evolution Cannot Create Biological Information. In *The Nature of Nature*, ed. Bruce Gordon and William A. Dembski. Wilmington, DE: ISI Books. Chapter 16.
Dembski, William A., Winston Ewert, and Robert Marks II. 2016. *Introduction to Evolutionary Informatics*. Singapore: World Scientific.
Dennett, Daniel C. 1995a. *Darwin's Dangerous Idea: Evolution and the Meanings of Life*. New York, NY: Simon & Schuster.
Dennett, Daniel. 1995b. Comment on Brian Goodwin, 'Biology is Just a Dance.' *Edge Conversations*. Edge.org. Available at https://www.edge.org/conversation/brian_goodwin-chapter-4-biology-is-just-a-dance#21968.
Denton, Michael. 2013. The Types: A Persistent Structuralist Challenge to Darwinian Pan-Selectionism. *Bio-Complexity* 3: 1–17.
———. 2016. *Evolution: Still a Theory in Crisis*. Seattle, WA: Discovery Institute Press.
Denton, Michael, and Craig Marshall. 2001. Laws of Form Revisited. *Nature* 410 (6827): 417.
Denton, M.J., Craig Marshall, and Michael Legge. 2002. The Protein Folds as Platonic Forms: New Support for the Pre-Darwinian Conception of Evolution by Natural Law. *Journal of Theoretical Biology* 219 (3): 325–342.

Draper, Paul. 2002. Irreducible Complexity and Darwinian Gradualism: A Reply to Michael J. Behe. *Faith and Philosophy* 19 (1): 3–21.

de Duve, Christian. 1984. *A Guided Tour of the Living Cell.* New York: Scientific American.

Erwin, Douglas H. 2016. *Wonderful Life* Revisited: Chance and Contingency in the Ediacaran-Cambrian Radiation. In *Chance in Evolution*, ed. Grant Ramsey and Charles H. Pence, 327–353. Chicago, IL: University of Chicago Press.

Ewert, Winston. 2016. Ev Ever Again. *Evolution News and Views*, March 23. Available at https://evolutionnews.org/2016/03/ev_ever_again/.

———. 2017. Digital Evolution: Predictions of Design. In *Theistic Evolution: A Scientific, Philosophical and Theological Critique*, ed. J.P. Moreland, Stephen C. Meyer, Christopher Shaw, Ann K. Gauger, and Wayne Grudem, 151–216. Wheaton, IL: Crossway.

Ewert, Winston, Dembski, William A., and Marks, Robert II. 2009. Evolutionary Synthesis of Nand Logic: Dissecting a Digital Organism. In *Proceedings of the 2009 IEEE International Conference on Systems, Man, and Cybernetics*, San Antonio, TX. October 2009, pp. 3047–3053.

Ferrera, Evandro, and Andreas Wagner. 2011. Evolutionary Innovations and the Organization of Protein Functions in Genotype Space. *PLoS ONE* 5 (11): e14172.

Fodor, Jerry, and Massimo Piattello-Palmarini. 2011. *What Darwin Got Wrong.* London: Profile Books.

Fortuna, Miguel A., Luis Zaman, Charles Ofria, and Andreas Wagner. 2017. The Genotype-Phenotype Map of an Evolving Digital Organism. *PLoS Computational Biology.* 13 (2): e1005414.

Freeman, Scott, and John C. Herron. 2007. *Evolutionary Analysis.* Upper Saddle River, NJ: Pearson Prentice Hall.

Fry, Iris. 2000. *The Emergence of Life on Earth: A Historical and Scientific Overview.* New Brunswick, NJ: Rutgers University Press.

Gauger, Ann, and Douglas Axe. 2011. The Evolutionary Accessibility of New Enzyme Functions: A Case Study from the Biotin Pathway. *Bio-Complexity* 1: 1–17.

Giberson, Karl, and Mariano Artigas. 2006. *Oracles of Science: Celebrity Scientists Versus God and Religion.* Oxford: Oxford University Press.

Gishlick, Alan. 2004. Evolutionary Paths to Irreducible Systems: The Avian Flight Apparatus. In *Why Intelligent Design Fails: The Scientific Critique of the New Creationism*, ed. Matt Young and Taner Edis, 58–71. Piscataway, NJ: Rutgers University Press.

Glass, David, and Mark McCartney. 2014. Explaining and Explaining Away in Science and Religion. *Theology and Science* 12 (4): 338–361.

Goodwin, Brian. 2001. *How the Leopard Changed Its Spots: The Evolution of Complexity. First Princeton Edition with New Preface.* Princeton, NJ: Princeton University Press.

Gould, Stephen Jay. 1989. *Wonderful Life: The Burgess Shale and the Nature of History.* New York: W.W. Norton.

———. 1996. *Full House: The Spread of Excellence from Plato to Darwin.* New York, NY: Harmony Books.

Gould, Stephen Jay, and Elisabeth S. Vrba. 1982. Exaptation: A Missing Term in the Science of Form. *Paleobiology* 8 (1): 4–15.

Gray, Asa. 1860. Natural Selection Not Inconsistent with Natural Theology. *Atlantic Monthly for July, August and October*, 1860. *Darwiniana: Essays and Reviews Pertaining to Darwinism.* New York: D. Appleton, 1888 [1874].

———. 1888a [1874]. Charles Darwin: Sketch Accompanying a Portrait in 'Nature'. In *Darwiniana*, 283–288. New York, NY: D. Appleton & Company.

———. 1888b [1874]. *Darwiniana: Essays and Reviews Pertaining to Darwinism.* New York: D. Appleton.

Griffiths, Paul E. 1996. Darwinism, Process Structuralism, and Natural Kinds. *Philosophy of Science* 63 (3): S1.

Häggström, Olle. 2007a. Intelligent Design and the NFL Theorems. *Biology & Philosophy* 22 (2): 217–230.

———. 2007b. Uniform Distribution Is a Probability Distribution. Unpublished Paper. Available at http://www.math.chalmers.se/~olleh/reply_to_Dembski.pdf.

Hecht, Michael, Das Aditi, Abigail Go, Luke H. Bradley, and Wei Yinan. 2004. De Novo Proteins from Designed Combinatorial Libraries. *Protein Science* 13 (7): 1711–1723.

Himma, Kenneth. 2005. "The Application-Conditions for Design Inferences: Why the Design Arguments Need the Help of Other Arguments for God's Existence." *International Journal for Philosophy of Religion* Vol. 57. No. 1 (2005). 1–33.

Hoggard Creegan, Nicole. 2013. *Animal Suffering and the Problem of Evil.* Oxford: Oxford University Press.

Hoyle, Fred. 1983. *The Intelligent Universe: A New View of Creation and Evolution.* London: Michael Joseph Limited.

Ilardo, Melissa, Markus Meringer, Stephen Freeland, Bakhtiyor Rasulev, and H. James I.I. Cleaves. 2015. Extraordinarily Adaptive Properties of the Genetically Encoded Amino Acids. *Scientific Reports* 5 (1): 9414.

Jiménez, Jose, Ramon Xulvi-Brunet, Gregory W. Campbell, Rebecca Turk-MacLeod, and Irene A. Chen. 2013. Comprehensive Experimental Fitness Landscape and Evolutionary Network for Small RNA. *Proceedings of the National Academy of Sciences – PNAS* 110 (37): 14984–14989.

John, J.D., D.S. Linden, G.S. Hornby, A. Rodriguez-Arroyo, S.E. Seufert, B. Blevins, and T. Greeling. 2004. Evolutionary Design of an X-Band Antenna for NASA's Space Technology 5 Mission. *IEEE Antennas and Propagation Society International Symposium* 3: 2313–2316.

Johnson, Curtis. 2015. *Darwin's Dice: The Idea of Chance in the Thought of Charles Darwin.* New York, NY & Oxford: Oxford University Press.

Kacar, Betül. 2016. Rolling the Dice Twice: Evolving Reconstructed Ancient Proteins in Extant Organisms. In *Chance in Evolution*, ed. Grant Ramsey and Charles H. Pence, 312–327. Chicago, IL: University of Chicago Press.

Kauffman, Stuart A. 1996. *The Origins of Order: Self-Organization and Selection in Evolution.* Oxford: Oxford University Press.

Kauffman, Stuart. 2000. *Investigations.* Oxford: Oxford University Press.

Kauffman, Stuart A. 2008. *Reinventing the Sacred: A New View of Science, Reason and Religion.* New York, NY: Basic Books.

———. 2013. Evolution beyond Newton, Darwin and Entailing Law: The Origin of Complexity in the Evolving Biosphere. In *Complexity and the Arrow of Time*, ed. C.H. Lineweaver, P. Davies, and M. Ruse, 162–190. Cambridge: Cambridge University Press.

Kojonen, Erkki Vesa Rope. 2016. *The Intelligent Design Debate and the Temptation of Scientism.* London: Routledge.

Koonin, Eugene V. 2007. The Cosmological Model of Eternal Inflation and the Transition from Chance to Biological Evolution in the History of Life. *Biology Direct* 2 (1): 15.

Koonin, Eugene, and Yuri A. Wolf. 2010. Constraints and Plasticity in Genome and Molecular-Phenome Evolution. *Nature Reviews Genetics* 11: 487–498.

Laland, Kevin, Tobias Uller, Marc Feldman, Kim Sterelny, Gerd B. Müller, Armin Moczek, Eva Odling-Smee Jablonka, Wray John, A. Gregory, Hopi E. Hoekstra, Douglas J. Futuyma, Richard E. Lenski, Trudy F.C. Mackay, Dolph Schluter, and Joan E. Strassmann. 2014. Does Evolutionary Theory Need a Rethink? *Nature* 514 (7521): 161–164.

Lenski, R.E., C. Ofria, R.T. Pennock, and C. Adami. 2003. The Evolutionary Origin of Complex Features. *Nature* 423 (6936): 139–144.

Lewens, Tim. 2007. Adaptation. In *The Cambridge Companion to Philosophy of Biology*, ed. David L. Hull, 1–21. Cambridge: Cambridge University Press.

Livingstone, David N. 2020. The Telos of Darwin's Troubling Metaphor: Darwin's Pigeons, Dennett's Cranes, Fodor's Granny. Paper for the Conjunctive Explanations in Science and Religion Conference, September 2–September 3, 2020.

Loikkanen, Juuso. 2019. *A Bridge Between Science and Theology? William A. Dembski's Theory of Intelligent Design*. Publications of the University of Eastern Finland. Dissertations in Education, Humanities, and Theology. No. 140.

Louis, Ard A. 2016. Contingency, Convergence and Hyper-astronomical Numbers in Biological Evolution. *Studies in History and Philosophy of Biology & Biomedical Science* 58: 107–116.

McGhee, George. 1999. *Theoretical Morphology: The Concept and Its Applications.* Columbia University Press.

———. 2011. *Convergent Evolution: Limited Forms Most Beautiful.* Cambridge, MA: The MIT Press.

McLeish, Tom. 2020. Evolution as the Unwrapping of the Gift of Freedom. *Scientia et Fides* 8 (2): 43–64. https://doi.org/10.12775/SetF.2020.014.

Meester, R.W.J. 2009. Simulation of Biological Evolution and the No Free Lunch Theorems. *Biology and Philosophy* 24: 461–472.

Meyer, Stephen C. 2013. *Darwin's Doubt: The Explosive Origin of Animal Life and the Case for Intelligent Design.* San Francisco, CA: HarperCollins.

Miller, Kenneth. 2002. *Finding Darwin's God: A Scientist's Search for Common Ground Between God and Evolution.* New York, NY: Harper Perennial.

Moreland, J.P., Stephen C. Meyer, Christopher Shaw, Ann K. Gauger, and Wayne Grudem. 2017. *Theistic Evolution: A Scientific, Philosophical and Theological Critique.* Wheaton, IL: Crossway.

Moritz, Joshua M. 2019. Contingency, Convergence, Constraints, and the Challenge from Theodicy in Creation's Evolution. In *God's Providence and Randomness in Nature: Scientific and Theological Perspectives*, ed. Robert J. Russell and Joshua M. Moritz, 289–328. West Conshohocken, PA: Templeton Press.

Conway Morris, Simon, and Stephen Jay Gould. 1998. Showdown on the Burgess Shale. *Natural History Magazine* 107 (10): 48–55. Available at http://www.stephenjaygould.org/library/naturalhistory_cambrian.html.

Musgrave, Ian. 2004. The Evolution of the Bacterial Flagellum. In *Why Intelligent Design Fails: The Scientific Critique of the New Creationism*, ed. Matt Young and Taner Edis, 72–84. Piscataway, NJ: Rutgers University Press.

Nilsson, Dan-e, and Susanne Pelger. 1994. A Pessimistic Estimate of the Time Required for an Eye to Evolve. *Proceedings of the Royal Society of London. Series B.* 256 (1345): 53–58.

Noble, Denis. 2008. *The Music of Life: Biology Beyond the Genome.* Oxford: Oxford University Press.

Orr, Allen. 1996. Darwin v. Intelligent Design (Again). Boston Review. December/January Issue.

Oslington, Paul. 2012. God and the Market: Adam Smith's Invisible Hand. *Journal of Business Ethics* 108 (4): 429–438.

Paley, William. 2008 [1802]. *Natural Theology.* Oxford World's Classics. Oxford: Oxford University.

Pallen, Mark J., and Nicholas J. Matzke. 2006. From the Origin of Species to the Origin of Bacterial Flagella. *Nature Reviews Microbiology* 4 (10): 784–790.

Perakh, Mark. 2004. There Is a Free Lunch After All: Dembski's Wrong Answers to Irrelevant Questions. In *Why Intelligent Design Fails: The Scientific Critique of the New Creationism*, ed. Matt Young and Taner Edis, 153–171. Piscataway, NJ: Rutgers University Press.

Pigliucci, Massimo. 2010. Genotype–Phenotype Mapping and the End of the 'Genes as Blueprint' Metaphor. *Philosophical Transactions of the Royal Society B* 365 (1540): 557–566.

———. 2012. Landscapes, Surfaces, and Morphospaces: What Are They Good For? In *The Adaptive Landscape in Evolutionary Biology*, ed. Ryan Calsbeek and Erik I. Svensson, 26–38. Oxford: Oxford University Press.

———. 2016.The Neo-Platonic Argument for Evolution Could not be More Wrong. *Nautilus*. http://nautil.us/blog/the-neo_platonic-argument-for-evolution-couldnt-be-more-wrong.

Pigliucci, Massimo, and Gerd B. Müller, eds. 2010. *Evolution: The Extended Synthesis*. Cambridge, MA: The MIT Press.

Pressman, Abe D., Ziwei Liu, Evan Janzen, Celia Blanco, Ulrich F. Müller, Gerald F. Joyce, Robert Pascal, and Irene Chen. 2019. Mapping a Systematic Ribozyme Fitness Landscape Reveals a Frustrated Evolutionary Network for Self-Aminoacylating RNA. *Journal of the American Chemical Society* 141: 6213–6223.

Ridley, Mark. 2004. *Evolution*. Oxford: Oxford University Press.

Riskin, Jessica. 2016. *The Restless Clock: A History of the Centuries-Long Argument over What Makes Living Things Tick*. Chicago: The University of Chicago Press.

Rockoff, Hugh. 2011. Parallel Journeys: Adam Smith and Milton Friedman on the Regulation of Banking. *Journal of Cultural Economy* 4 (3): 255–283.

Ruse, Michael. 2013. Wrestling with Biological Complexity: From Darwin to Dawkins. In *Complexity and the Arrow of Time*, ed. C.H. Lineweaver, P. Davies, and M. Ruse, 279–307. Cambridge: Cambridge University Press.

Sedley, David. 2007. *Creationism and Its Critics in Antiquity*. Sather Classical Lectures 66. Berkeley, CA: University of California Press.

Sepkoski, David. 2008. Macroevolution. In *The Oxford Handbook of Philosophy of Biology*, ed. Michael Ruse, 211–237. Oxford: Oxford University Press.

Shanks, Niall. 2007. *God, the Devil and Darwin: A Critique of Intelligent Design Theory*. Oxford: Oxford University Press.

Simon, Dan. 2013. *Evolutionary Optimization Algorithms: Biologically-Inspired and Population-Based Approaches to Computer Intelligence*. Hoboken, NJ: John Wiley & Sons.

Sober, Elliott. 2008. *Evidence and Evolution: The Logic Behind the Science*. Cambridge: Cambridge University Press.

Taylor, S.V., K.U. Walter, P. Kast, and D. Hilvert. 2001. Searching Sequence Space for Protein Catalysts. *Proceedings of the National Academy of the Sciences USA* 98: 1056–1060.

Tian, Pengfei, and Robert B. Best. 2017. How Many Protein Sequences Fold to a Given Structure? A Coevolutionary Analysis. *Biophysics Journal* 113 (8): 1719–1730.

Tokuriki, Nobuhiko, and Dan S. Tawfik. 2009. Stability Effects of Mutations and Protein Evolvability. *Current Opinion in Structural Biology* 19 (5): 596–604.

Venkatarayalu, N. V., and Ray, T. 2003. Single and Multi-Objective Design of Yagi-Uda Antennas Using Computational Intelligence. *The 2003 Congress on Evolutionary Computation*, CEC '03, 8–12 December 2003. 1237–1242.

Wagner, Andreas. 2014. *Arrival of the Fittest: Solving Evolution's Greatest Mystery.* New York, NY: Current.

———. 2016. Without a Platonic Library of Forms, Evolution Couldn't Work. *Aeon.com.* March 16. Available at https://aeon.co/essays/without-a-library-of-platonic-forms-evolution-couldn-t-work.

Ward, Keith. 2004. Theistic Evolution. In *Debating Design: From Darwin to DNA*, ed. William A. Dembski and Michael Ruse, 261–274. Cambridge: Cambridge University Press.

Weinreich, Daniel, Nigel F. Delaney, Mark A. DePristo, and Daniel L. Hartl. 2006. Darwinian Evolution Can Follow Only Very Few Mutational Paths to Fitter Proteins. *Science* 312 (5770): 111–114.

Wichmann, Stefan, and Zachary Ardern. 2019. Optimality in the Standard Genetic Code Is Robust with Respect to Comparison Code Sets. *Bio Systems* 185: 104023.

Willmer, Patricia. 2003. Convergence and Homoplasy in the Evolution of Organismal Form. In *Origination of Organismal Form: Beyond the Gene in Developmental and Evolutionary Biology*, ed. Gerd Müller and Stuart Newman, 33–50. Cambridge, MA: The MIT Press.

Wolpert, D. H. 2002. William Dembski's Treatment of No Free Lunch Theorems Is Written In Jello. Posted on December 13, 2002 at *Talk Reason.* Available at www.talkreason.org/articles/jello.cfm.

Wolpert, D.H., and W.G. Macready. 1997. No Free Lunch Theorems for Optimization. *IEEE Transactions on Evolutionary Computation* 1 (1): 67–82.

CHAPTER 5

Indirect Design and the Revelatory Potential of Biology

As I have pointed out in Chap. 2, for Dawkins, Darwinian evolution functions as a "consciousness-raiser" which shows that apparently convincing examples of purposeful order can in fact be explained naturalistically. For Dawkins, since Darwinism shows that we can explain design without a designer, this provides grounds for similarly explaining all putative examples of divine design (including fine-tuning) by reference to non-purposeful causes. Thus, "a deep understanding of Darwinism teaches us to be wary of the easy assumption that design is the only alternative to chance, and teaches us to seek out graded ramps of slowly increasing complexity. [...] After Darwin, we should feel, deep in our bones, suspicious of the very idea of design" (Dawkins 2006, 139).

However, considered in the light of evolution's dependence on its preconditions, explored in the previous chapter, does evolution indeed show what Dawkins claims? Suppose for the sake of the argument that a divine designer is actually responsible for the laws of form (and other environmental factors) that enable evolution. In that case, evolution would be dependent on design, and therefore evolution would not actually show us that evolution can produce design without a designer. Therefore, to use the non-designedness of evolution as a basis for criticizing design arguments more broadly, we would first have to establish that the wider teleology that evolution depends on is also not designed. Otherwise we would be assuming the very thing that we are trying to prove, and we would be

E. V. R. Kojonen, *The Compatibility of Evolution and Design*, Palgrave Frontiers in Philosophy of Religion,
https://doi.org/10.1007/978-3-030-69683-2_5

arguing in a circle. It seems clear, then, that evolution cannot function as a premise to ground this kind of general suspicion of design arguments.

At least in this way, then, knowledge of the wider teleology required for evolution seems to undermine evolution as an argument against design. As others have also argued, and as I have argued in Chap. 2, evolution is at least compatible with belief in divine purpose. But could we say more, and state that evolution is also compatible with interpreting biological order as revealing signs or evidence of the Creator?

In this chapter, I will analyze the conditions for combining either design discourse or a biological design argument with Darwinian evolution, and how this compatibility might be defended. Again, the general compatibility of evolution and design is not at issue here; what is at issue is the question of the revelatory potential of the biological creation. I am looking for ways to harmonize design and evolution as explanations, where design still retains some basis in experience, so that such a reading of nature is not based solely on one's religious background assumptions, but finds some resonance with the evidence. I will begin with the general conditions of how and when design and natural explanations might conflict or be compatible. After this, I will analyze previous attempts at defending the language of biological design within a theistic evolutionist framework. Having developed my own account, I will then argue for it based on thought experiments.

5.1 Design and Natural Explanations

Compatibility or Incompatibility?

It is not difficult to find cases in which designers actually work by setting up processes and mechanisms that then produce the desired end result without the need for further intervention by the designer. In the previous chapter, I mentioned genetic algorithms as an example of this, but any use of mechanisms and automation would also suffice to make this general point. The compatibility of natural causes and design was already asserted by Plato, who argued that the Creator used the existing properties of matter in his work, and that intentional and material explanations can be compatible. For instance, while the poison drunk by Socrates was the material cause of his death, this does not mean that the politics of Athens and the character of Socrates have no explanatory role (Sedley 2007, chapter 4).

Nor did Paley think that the order of biological organisms must be directly created by God in order for design to be evident. Supposing that his "watch on a heath" had the ability to reproduce, to give birth to another watch, this would, in Paley's opinion, only increase our admiration for the watchmaker. At this point it would "now no longer [be] probable, that the individual watch, which our observer had found, was made immediately by the hand of an artificer, yet doth not this alteration in anywise affect the inference, that an artificer had been originally employed and concerned in the production. [...] All these properties, therefore, are as much unaccounted for, as they were before" (Paley 2008 [1802], 11–12). For Paley, the natural explanation only moves the problem back rather than solving it. The process of reproduction is about the transmission of teleological properties rather than the ultimate origin of these properties.

As I have noted (Kojonen 2016a, 169–181), proponents of ID are similarly not opposed to the general idea that the designer, or God, could have worked through natural laws and mechanisms. As Behe (1999) writes, the alternatives are not that of "a cartoon world, where genies and fairies swirl about endlessly dispensing magic, or a world of relentless materialism where, say, the charitable work of a Mother Teresa is explained only in terms of evolutionary selection coefficients." Proponents of ID thus think of design and natural causes as generally complementary, and even argue that it is possible to define "theistic evolution" in a way that is compatible with their design arguments, namely as a process directed in a tangible, detectable way by a designer. But this, according to proponents of ID, is not compatible with mainstream understandings of Darwinian evolution. For example, according to Johnson (1993), "'Evolution,' honestly understood, is not just a gradual process of development that a purposeful Creator might have chosen to employ. It is, by Darwinist definition, a purposeless and undirected process that produced mankind accidentally." Dembski (1999, 111–112) similarly writes that "For the hard-core naturalist, theistic evolution at best includes God as an unnecessary rider in an otherwise purely naturalistic account of life."

Dembski, in defending ID, sometimes comes close to admitting the compatibility of evolution and design. He argues, for example, that "Intelligent design is not a theory about the frequency or locality or modality by which a designing intelligence intervenes in the material world. It is not an interventionist theory at all. Indeed, intelligent design is perfectly compatible with all the design in the world coming to expression by the ordinary means of secondary causes over the course of natural history" (Dembski 2008, 221). But Dembski and other ID proponents nevertheless see evolution by natural selection and mutation as a type of secondary causation that is not compatible with design. This is because, they argue, such evolutionary processes are by definition unguided, being based on random variations, nor do they leave a sufficient causal role for the designer. This, then, means that the designer's existence will be undetectable. Meyer similarly argues that if theistic evolutionists do not ascribe a role to God beyond establishing the initial conditions and maintaining the laws that make evolution possible, then theistic evolution "seems to be little more than an empty phrase describing a theory that, in the end, reduces to a standard, and completely inadequate, materialistic mode of explanation for the origin of biological information" (Meyer 2017, 236).

In *The Intelligent Design Debate and the Temptation of Scientism* (Kojonen 2016a, chapter 169–189), I argue that such an understanding of evolution gives too much weight to a naturalistic philosophical interpretation of evolution, bypassing the possible grounds of a separation between evolutionary biology and this philosophical interpretation (no matter how popular the philosophical interpretation has historically been). But here I will focus on the question of the compatibility of design and evolution as explanations. As with the critiques of the compatibility of design-based and Darwinian explanations analyzed in the previous chapter, the core issue here is the alleged unnecessity of a designer, once Darwinian explanations are accepted. The principle of parsimony, colloquially known as Occam's razor, is crucial here. Simplicity is widely understood as a virtue of explanations, meaning that if we want to use a more complex explanation involving several factors (such as both design and natural causes), then we need to justify this by added explanatory power. In other words, if evolution explains biological teleology adequately, then we should not appeal to other explanatory factors. In the philosophy of mind, a similar argument has been famously defended by Kim (2007), who argues against explanations involving *causal overdetermination*. The idea is that if one explanation is sufficient, then it would be epistemically virtuous not to appeal to multiple explanations.[1]

However, as Glass and Schupbach (2020) note, simplicity is not the only criterion of evaluating explanations. In *conjunctive explanations*, several different explanations are used together to explain various facets of a complex phenomenon, together providing a more complete explanation that justifies the added complexity. Here, then, we once again return to the argument that evolution makes divine design an unnecessary explanation for biological teleology, and how design might still be a justified part of explaining biological teleology even in an evolutionary cosmos. On the one hand, the examples of Socrates' death and Paley's watch already make it clear that there are cases in which design and natural explanations are indeed complementary. On the other hand, surely there are also cases where one explanation makes another unnecessary. To delve deeper into the matter, then, I will discuss how we should conceptualize the relationship of these different explanations in general, and consider what sets apart cases of compatibility and incompatibility. As with the earlier discussion of thought experiments, it is also crucial here that our account of compatibility or incompatibility can deal with actual examples, as well as with thought experiments. And in order to avoid ruling out either design or non-design based on a priori considerations, it is desirable to be able to point to at least some cases of compatibility and incompatibility—and then check which the case of evolution resembles.

Developing an Account of Compatibility

Philosophers have proposed many different ways of conceptualizing the relationship of different explanations. One popular way is by using the spatial metaphor of "levels of explanation," assuming an account of the stratification of reality, where the methods suited for studying one level may not fit all levels (McGrath 2002, chapter 10; Bhaskar 1978). Another is by adapting the distinction between proximate and ultimate explanations, first popularized in the philosophy of biology by Mayr (1961).[2] In Mayr's parlance, proximate explanations refer to the operation of the biological organism as it stands, ultimate explanations to the historical causes that are the origins of this present-day organization. Here, the proximate level roughly corresponds to the question "How does the organism operate?" and the ultimate level to the question "Why/How did it come to operate in this way?" While this distinction has been criticized for not capturing the full complexity of biological causation, it nevertheless continues to be widely used as a rough approximation (Laland et al. 2011; Dickins and Barton 2013). Applied to design arguments, however, evolution would stand as the proximate historical explanation, while divine design stands as the ultimate explanation that ultimately works through evolution. Here the distinction is similar to St Thomas' primary causation and secondary causation, where natural beings, being endowed with their own causality, are nevertheless dependent on God as the primary cause which sustains all in being (Feser 2017).

Using the metaphor of levels of explanation, Ratzsch and Koperski (2019) analyze what they call "level-shifting," arguing that design can work through natural causes. For example, suppose that an elderly uncle dies in suspicious circumstances, and relatives suspect that the niece killed her uncle. Police investigations, however, reveal a natural cause for the death: an overdose of the uncle's medication. Nevertheless, it may still be plausible to claim that the niece killed the uncle by mixing up his medication, so that the actions of an agent act as the distal or ultimate cause of death, working through the proximate natural cause. Similar examples could be multiplied. For example: Why did the window break? Because a rock flew into it. Why did the rock fly into it? Because it was thrown. In such examples, it seems clear that the ascription of a natural cause to an event does not necessarily decrease the responsibility of the human agent for the event. Rather, the causal chain is transitive, transferring the responsibility back to the agent.

However, Ratzsch and Koperski also point out a non-plausible example of level-shifting. Suppose that crop circles (which some UFO enthusiasts suppose are produced by aliens) are proven with video evidence to be made by humans. In 1991, in Britain, supposed crop circle experts were indeed fooled by two jokesters, who made such circles during the night and videotaped the process (Schmidt 1991). When the resultant crop circles were declared to be clearly the products of an alien intelligence, the jokesters were able to step forward with embarrassing proof to the contrary. A UFO enthusiast could respond by claiming that the aliens could have been mind-controlling the humans, and thus the aliens could still be responsible for the resulting pattern. However, here level-shifting is clearly implausible, because the alternative explanation eliminates the reason why the aliens were invoked as the explanation in the first place, and thus makes the reference to alien design unnecessary. Granted, in this case the alien designers are replaced with human designers, but we could in principle also imagine having video evidence of the crop circles being caused by a storm or other natural cause, in which case all references to designers would be unnecessary. Similarly, we can also imagine cases where instead of a stone being thrown by an agent, the stone is thrown by a powerful storm, eliminating the need to invoke an agent.

Another way to combine different explanations, without relying on the metaphor of levels or the proximate—ultimate distinction, would be to argue that different explanations are answers to different questions. In the philosophy of explanations, one currently popular account states that causal explanations should be understood as *contrastive*, explaining why this rather than something else happened (Persson and Ylikoski ed. 2007). For example, in the case of the poisoning of Socrates, we are asking two different though related questions, such as (1) "What is the physiological cause for the cessation of Socrates' heartbeat (why did it stop, rather than continue beating)?" and (2) "What psychological and moral reasons did Socrates have for drinking the poison that caused his death (why did he drink the poison, rather than not)?" Accepting this understanding of the nature of explanations as answers to questions, the compatibility of evolutionary explanations and design-based explanations for life also depends on what kind of questions both explanations are thought to answer. However, the levels metaphor and the proximate—ultimate distinction have the benefit of making the relationship between different explanations intuitively clear, since in these models it is clear that one explanation can be more fundamental, and can stand behind the other one in the causal chain of events.

When, then, should we think that two explanations as answers to some question overlap so much that having one explanation makes the other unnecessary? Glass uses the concept of "explaining away" to analyze further the elimination of one explanation over another. Whether something can be explained away depends on whether we have access to a well-established hypothesis that can explain all of the evidence. As Glass (2012, 79) notes in regard to design arguments:

> In most cases design is compatible with the alternative explanation, but if this is so, why not accept both design and the alternative explanation? The obvious answer is that there is no need to infer two explanations when one will do. When I learn that my children were playing in the study, the hypothesis that there has been a burglary becomes redundant as an explanation for the untidiness. In general, however, there is a question as to when one explanation is good enough to render the other redundant.

Applied to the question of design in biology, the question is: Does explaining biological adaptive complexity by reference to natural evolutionary mechanisms leave any grounds for design as an explanation, or is the evidence of design explained away? If alternative explanations more plausibly explain away the data that were originally the basis of the design argument, then they can in principle make design redundant. Schupbach (2016, 266) further elaborates on this criterion, arguing that explaining away would require that "one of the hypotheses in question, taken by itself, renders the other either irrelevant or negatively relevant to *E* [the evidence]. The most obvious (though not the only) way to pursue this line is to argue that the *D* or *N* in question suffices for a full explanation of *E* (implies *E*)." Here *D* and *N* stand for ***design*** and ***natural selection***, acting as explanations for the evidence (*E*).

And here is the proposed difference between Darwinian evolution and natural explanations which simply transfer the responsibility up the causal chain. Evolutionary explanations are commonly understood to explain teleology (or the appearance of teleology) itself via reference to a non-teleological process. Evolution is understood as an attempt to reduce teleology to non-teleological causes, and in this way explain the very same evidence that was given as grounds for inferring design. Thus, once we already have an explanation for biological teleology by way of Darwinism, it can be argued that we no longer have a need for any further explanations. The claim is that the question "What processes are responsible for the apparent teleology of biological nature?" has already been definitively answered by the theory of evolution, with no need for further explanations.

However, in the previous chapter I have argued that evolution pushes the explanation for life's order back to the laws of nature at least to some degree, and probably to a large degree. As the understanding of evolution in the biological sciences has deepened, the role of "laws of form" in explaining biological teleology has increased, and the efficacy of the process of natural selection and mutation has been recognized to depend on further factors. For example, if we accept the idea of a platonic library of forms that makes evolution possible, it seems that evolution no longer explains the forms themselves, but only their actualization. In Wagner's (2016) words, before evolution, the forms "already exist in a world of concepts, the kind of abstract concepts that mathematicians explore." If evolution is understood in this way, then it seems to me that Darwinism does not in fact answer the same question, at the same level, as the hypothesis of purposeful design. Something about the question of how such order can exist in our universe remains unanswered just by the appeal to known evolutionary mechanisms—even if no interventions by a designer within the cosmos are required for the explanation.

With this understanding of biological evolution, we can distinguish between two questions, a proximate and ultimate question regarding biological teleology. First (the proximate level), the question answered by Darwinism is related to the particular examples of biological teleology: Why does this particular organism have this teleological order in this particular historical situation, rather than not evolving—or evolving another strategy? And on a more general level, what are the historical processes that produce the appearance of teleology in this particular case? Second (the ultimate level), the question answered by the biological design argument would be something like "How is it possible for there to be biological teleology at all, even complex biological teleology, rather than not?" The idea is that it seems much more credible that a purely physical process would be able to produce such results, if the process has been designed to produce results of this type, than if the process had no design behind it at all. Thus, in an evolutionary view of design God would not necessarily be responsible for designing the particulars of organisms like a human artisan might fashion "the hinge of a door," to use Darwin's (1958 [1887], 87) phrase. However, God can still be responsible for the existence of biological teleology as the one who makes the whole Darwinian process possible, and who uses it to produce biological organisms according to His intentions (which might also allow for some real contingency within the processes). In effect, the evolutionary processes are caused by God, and hence God is also responsible for the products of evolution, through the transitivity of causes, just as in the previously described examples where design and natural causes are linked.[3]

This way of construing the difference between the Darwinian explanation, as the proximate explanation, and divine design, as the ultimate explanation for biological teleology, comes close to the perspective of theistic evolution outlined by Tracy (2013, 460), that "God as creator establishes the role of chance in the processes of nature, and thereby shapes the array of possibilities explored in the unfolding history of the created world." Nevertheless, nothing about this way of stating the question requires that this is all that God does. In addition to setting up the process of evolution, God could well also be more intimately involved in directing the process, and all outcomes of the process could well be specifically intentioned by God, because of his foreknowledge of what results the contingent process will create. Moreover, nothing about the design argument as such prevents even divine interventions in the process. The point is, though, that this invisible guidance and interventions do not play any necessary role in the design argument. God does not need to be directly responsible for every single detail of biological teleology, or to be the sole cause acting in order for divine design to be a central causal factor that explains the existence of biological teleology—a factor that is not at all causally irrelevant or peripheral to understanding why biological teleology exists.

On this account, it is not just that evolution makes it possible to believe in a designer for independent religious reasons. Rather, the evidence of biological teleology itself continues to be suggestive of design despite the reality of evolution. On this account, evolution does not make design into an unnecessary explanation, because evolution pushes the problem back to the wider teleology of the cosmos and the "laws of form." Because of evolution's dependence on tight preconditions, it does not provide a self-sufficient "third option" for explaining teleology that would sit between chance and design. This was also the reason why Asa Gray thought that evolution does not in any way diminish the force of the biological design argument. Though Gray may have been wrong to suggest guided variations as a theistic addendum to evolutionary biology, it seems to me that he was nevertheless right to conclude that here evolution "leaves the question of design just where it was before [...] the issue between the skeptic and the theist is only the old one, long ago argued out—namely, whether organic Nature is a result of design or of chance" (Gray 1860). However, in contrast to the ID movement's approach, this is a philosophical difference between the theist and the naturalist, not a scientific dispute.

As Glass and Schupbach (2020) note, there is always an explanatory cost to conjunctive explanation, as such an explanation will inevitably be more complex than just appealing to a single explaining factor. But as long as the conjunctive explanation succeeds in increasing explanatory power, such as by responding to more explanatory questions, and makes the result less surprising, then the more complex explanation can be justified. After all, Occam's razor is not the only criterion of a successful explanation. It seems to me that, based on the above, the theistic explanation does clearly add to the understanding of biological teleology, since there is an explanatory question that is not yet answered by evolutionary explanations alone, particularly because evolutionary explanations partly push back the explanation to the laws of form. However, the explanatory cost of this explanation will of course be affected by how costly theism is thought to be. If we already believe in a Creator, for reasons such as those hinted at in Chap. 2, then there will be little explanatory cost in referring to divine design as an essential part of the complete explanation for biological teleology. However, if we think that there is much reason not to believe in a Creator, then this will of course make the change required by the biological design argument much larger, and thus perhaps more difficult to make.

As noted, Gray envisioned divine action as making a difference for evolution in two main respects: (1) being the ultimate cause working through the process and (2) guiding the apparently random variations to ensure particular outcomes. Gray himself did not see the guidance of mutations as crucial for his account, and as noted, the explanatory value of divinely guided mutations has been decisively rejected by mainstream evolutionary biology ever since Darwin. I have also argued that the first way is sufficient for the biological design argument, and in the rest of this chapter my arguments will also follow this line. As noted, this line of argument is supported by the idea that evolution pushes back the problem of explaining biological teleology at least to some extent.

Nevertheless, against views which would emphasize the contingency and improbability of evolutionary outcomes to a great extent, some have presented an alternative response relying on still giving guided mutations an explanatory role in a theistic, non-scientific interpretation of evolutionary theory. As noted in Chap. 4, some theologians and philosophers see the apparent directionality of evolution as evidence for purpose behind the process. Glass and McCartney (2014, 348–350) argue that if such directionality in evolution is seen as very improbable and contingent (following a Gouldian understanding taken to the extreme), then explaining the existence of such directionality despite its improbability may well allow for an explanatory role for divine guidance, through the divine guidance of mutations or other events affecting evolutionary history. Glass and McCartney quote Brooke's (1991, 315) summary of the Princeton theologian James McCosh (1811–1894), who also made a similar claim, though related more to the outcome of evolution than the overall directionality of the process: "If the human race were the product of so many contingencies, then either it was a more monumental fluke than atheists had ever dreamed, or, as McCosh wryly put it, the prevalence of accident could not be accidental." Glass and McCartney's argument allows for such explanatorily relevant divine guidance of evolution to happen in a number of ways, including the guidance of the starting conditions, environments, and mutations that influence evolution. This would then allow for divine guidance to have a role in a conjunctive explanation of evolution. However, it seems to me that it would be desirable to find a way to defend the explanatory relevance of design without requiring elements like guided mutations that would substantially differ from the mainstream understanding. Moreover, it seems to me that the biological evidence supports the idea that the environment itself, with its laws of form and other factors, provides some directionality to the evolutionary process. Therefore, my goal is to reconstruct the biological design argument in a way consistent with such an understanding of evolution—without requiring that evolutionary explanations are improbable in the context of the questions that these explanations are supposed to answer.[4]

I have now presented a preliminary sketch of how I understand the compatibility of the biological design argument and evolution. However, there are still questions to be answered, and further arguments to be presented in support of this basic model. For example, I have presented the compatibility of design and evolution as a possible defense on behalf of the "theist on the street," who regards the intuitive perception of design in biology as evidence supporting religious belief. But my own defense utilizes arguments in molecular and theoretical biology, which are going to seem fairly arcane to the theist on the street, and whose interpretation is still in question even among practicing biologists. Thus, while I have now presented arguments showing that the biological design argument as an argument still works, there is still need to examine how this might affect the intuitions about design. Many such intuitive theists are likely to hold that God is directly responsible for designing the details of biological organisms. In such cases, it might be claimed that this account of the compatibility of biological design and Darwinian evolution does not really save the validity of the intuitions of the "theists on the street", since these theists would be wrong about the manner in which God brought about the existence of biological teleology. Moreover, the relationship between fine-tuning arguments and this design argument also needs to be examined further, as does the discussion of the problem of evil and bad design. If God is responsible for evolution, then is God not responsible for these as well?

To respond to these (and other) questions, I will turn to analyzing strategies that have been used by other defenders of the compatibility of design and evolution, and see what insights might be gleaned from them. I will then also turn to defending the account in light of thought experiments. After this, I will turn to the problem of evil.

5.2 From Possibility to Plausibility

Defending Design Discourse

As noted, despite his critique of Paley's design argument, Darwin continued to believe that the idea of design has intuitive force. His conversation with the Duke of Argyll (Argyll 1885) shows this well:

> In the course of that conversation I said to Mr. Darwin, with reference to some of his own remarkable works on the *Fertilisation of Orchids*, and upon *The Earthworms*, and various other observations he made of the wonderful contrivances for certain purposes in nature—I said it was impossible to look at these without seeing that they were the effect and the expression of Mind. I shall never forget Mr. Darwin's answer. He looked at me very hard and said, 'Well, that often comes over me with overwhelming force; but at other times,' and he shook his head vaguely, adding, 'it seems to go away.'

As I have noted in Chap. 2, Plantinga's (2011, chapter 8) account of design discourse builds on the ubiquity of this kind of intuitive perception of design in nature. He claims that rational design beliefs can arise non-inferentially, akin to perceptual beliefs. Others can also adopt the same strategy and argue for the rationality of taking our intuitive design beliefs about biology as rational until proven otherwise. As will be clear by this point, I am sympathetic to Plantinga's goal of defending the rationality of ordinary intuitive beliefs in design. However, evolution is a possible undercutting defeater for these beliefs. There is a need to examine whether Plantinga's response is sufficient.

Plantinga's claim is that in purported examples of conflict between science and religion, the conflict is typically actually between naturalistic philosophy and religion, rather than between science and religion. Thus, for example, the problem with evolution is not evolution as such, but evolution interpreted through the lens of naturalistic philosophy. Since, in this understanding, evolutionary science itself does not rule out divine guidance, it is, Plantinga (2011, chapters 1, 2 & 8) argues, consistent with the beliefs produced by intuitive design discourse. However, granting the general compatibility of evolution and design does not yet respond to evolution as an undercutting defeater, a defeater which removes the grounds of the belief. It seems that design discourse includes the intuition that biological teleology cannot be explained without design. If evolutionary biology shows that these features are plausibly explained without any need for intelligent guidance, then this should be relevant for the rationality of the believer's beliefs which are grounded in these features.

Plantinga argues that this undercutting defeater does not work for two reasons. First, he argues that evolution can at most give us an improbable explanation of how the appearance of design might be explained without design. However, demonstrating that something is possible does not yet demonstrate that it actually happened. For instance, it is possible that Plantinga has won the lottery, but this does not yet give us reason to believe that this actually happened. For Plantinga, the epistemic status of evolution is akin to such an improbable possibility, which is not sufficiently probable for it to override the reliability of our design detection in biology. Here, as also in some of his previous discussions of faith and science, Plantinga relies on the ID movement's arguments, particularly Behe's irreducible complexity, to buttress his argument (Plantinga 2011, 225–264; Dennett and Plantinga 2011).

However, as I have noted (Kojonen 2016b, 378; Kojonen 2018), this argument is unsatisfactory as a harmonization of Darwinism and design because Darwinian evolutionary biology is not meant to demonstrate merely that the evolution of life by natural mechanisms is not astronomically improbable. Rather, it is argued that in our kind of cosmos, it is probable for evolution to occur, and we have good evidence in favor of the evolutionary explanation for life's development, such as the evidence for common descent. Plantinga is correct that evolutionary theory is still incomplete. But to argue that evolution is not a probable explanation of life's complexity is to side with ID against evolutionary theory. Whatever the merits of such an approach, at least it cannot be said to truly harmonize belief in Darwinian evolution and design. While evolution is accepted as a highly improbable possibility, design, not evolution, is still the explanation that is believed to be true (Plantinga 2011, 256).[5]

Plantinga's second response is to argue that the theist also has other reasons to believe that evolution was guided by God. According to Plantinga (2011, 261), even if the evolutionary explanation seemed to replace design as the explanation considered solely on the merits of the scientific arguments, we could still have reason to affirm design from the perspective of faith. Traditional theists affirm that God can act in the world invisibly, and believe that God is Creator. Thus, Plantinga argues that the theist knows that teleology was actually intentionally created by God in a process of guided evolution. Therefore, theists can rationally believe, as part of their Christian faith, that the interpretation of biology as designed is correct, despite the seeming possibility of an evolutionary explanation.

However, this defense of design discourse leaves the recognition of biological order as a sign of the Creator as something to be recognized only as part of a broader religious worldview, leaving the evolutionary objection intact for all others (Kojonen 2018, 448). It does not provide a response to evolution as an undercutting defeater, just a way for the theist to possibly sidestep this defeater. Perhaps more pressingly, this defense also does not leave any reason to think that biology should function as a sign of the Creator any more than anything else in creation. After all, as part of a theology of nature, religious believers rationally believe that all of nature, not just biological teleology, is created by God. Why, then, should biological teleology function as any sort of special sign of the Creator, that would warrant engaging in "design discourse?" Thus, the strategy of appealing to a theology of nature to defend design discourse threatens to undermine the point of engaging in any design discourse.

The example of crop circles is relevant here. In the example, the basis for positing aliens as the explanation for crop circles is lost because the alien hypothesis and the human hypothesis explain the same data, making the aliens unnecessary. Nevertheless, it remains possible to harmonize the alien hypothesis with the video evidence. It is possible that the tape is fake or misinterpreted (just as the evidence for evolution could be misinterpreted and misrepresented), or that the humans were being mind-controlled by aliens (as the evolutionary process could be directed by God). Someone firmly committed to the alien hypothesis could still argue that their broader worldview makes the alien hypothesis as an explanation of at least some crop circles credible for them. However, this would be irrational to believe for those who do not share the worldview, and think there is reason to disbelieve in alien visitations to Earth. Nor would the evidence of crop circles then add any evidential force even to those who already believe in alien visitations based on other grounds. My worry is that Plantinga's argument will leave Christian believers in the same position regarding evolution as the hypothetical UFO believer in the case of crop circles. The Christian believer can have other very good reasons for thinking that Christianity is true (or they could just believe this as a basic belief, following Plantinga), but the designedness of biological organisms would not be one of those reasons.

There is a need to flesh out Plantinga's model by arguing somehow that the apparent design in biology really is a reliable sign of design at least from the Christian religious point of view, and to explain how the information provided by the Christian tradition might enable perceiving design in biology, rather than simply distorting our perspective. Recall Lamont's (2004, 12) point: "No enhancement of cognitive capacities can enable them to go beyond what the evidence available to them can warrant them in believing; if a change to our cognitive capacities were to do this, it would not be an enhancement but a corruption of them." Thus, following the previous sections, my own suggestion would be to defend the rationality of design discourse by arguing that evolutionary explanations do not actually undercut design-based explanations of biological teleology. My suggestion is that Plantinga could defend design discourse by arguing that evolution, even if accepted, still leaves something about biology to be explained, and that this something is the very thing that triggers our design perceptions in the first place. This would then create a crucial difference between design discourse and the crop circle example. Design would not just be "read into" nature, but also "read out" of nature.

Mats Wahlberg presents ideas that are useful for this strategy. According to Wahlberg (2012, 182), "Theism entails that "God's immense intelligence and intentional activity are *necessary conditions* for the emergence of complex organisms. From this it follows that it requires immense intelligence to bring it about that the property of being a complex organism is instantiated." Just like Plantinga, Wahlberg claims that the beliefs of theists make it reasonable for them to regard their design perceptions as veridical. However, in Wahlberg's argument, there is a clearer understanding of how the theist's design beliefs could actually be grounded in reality, rather than being just part of human psychology. The properties of biological organisms are actually expressive of design, and the theist also accepts a general principle that God's intentional activity is a required precondition for the emergence of the existence of complex organisms. Thus, Wahlberg argues, the theist's additional knowledge undercuts the assertion that evolution could ever produce design without a designer. If all depends on God, then so does evolution, and the perception of design in biology is not in error—at least from the point of view of the theist.

However, it seems that there is still room to formulate more specific principles. This is because for theists, God's existence is actually the necessary condition for the existence of anything at all, not just biological organisms. As in the cosmological argument, which Wahlberg himself defends elsewhere, the existence of anything that is contingent requires God as the necessary ground of being (Wahlberg 2017). Thus, everything in nature has properties that are "intrinsically connected to a creator in the sense that their instantiation presupposes the existence of a creator," to reuse Wahlberg's phrasing. Why, then, should we think that there is something special about biology that gives grounds for the perception of design, as opposed to ordinary rocks, which also require the existence of the Creator? Is it not then just a matter of human psychology that we tend to see biological teleology as a special sign of the Creator, rather than perceiving the Creator equally through everything that exists? In defense of Plantinga and Wahlberg, part of the point of design discourse is to claim that rational belief in design does not require being able to elucidate what, exactly, about the appearance of teleology in biology causes us to perceive it as designed. Rather, the point is that this is non-inferential basic knowledge, and some properties are just especially expressive of the Creator.

Nevertheless, it does still seem that it is desirable to be able to defend the rationality of design discourse further. And in defending the especially expressive nature of biological order, it seems to me that defenders of design discourse will need to appeal to similar grounds as the defenders of design arguments. Wahlberg's previously quoted theistic defense of design perception builds on the basic principle that when something (A) is intrinsically, necessarily connected to something else (B), then A implies B. The same also works if the connection is not necessary, for in that case we can then at least say that A gives evidence supporting belief in B. Hence, if the defender of design discourse can show some kind of intrinsic or probabilistic connection between complex teleology and design, then this will help show why such properties are especially expressive of the divine mind. Darwinian evolution might be brought forward as a counterargument to the proposed experiential link between complex teleology and design. I have argued that this counterargument is not convincing, since evolution and design can both be supported by the evidence. However, here theistic background beliefs, as formulated by Wahlberg, can provide further support from the point of view of the theist. If the theist has additional information that evolution also depends on God, then it becomes even clearer for the theist that evolution does not work as a rebuttal of the link between teleology and design.

Even for nontheists, it should be clear that, at the very least, not just any universe can create life, and life's complex teleology can act as a sign that our universe is very special indeed. As Wahlberg (2012, 182) argues, the proposition that "it requires immense intelligence to bring it about that the property of being a complex organism is instantiated [...] is highly plausible. It *seems* to be very hard to create, from non-living matter, the highly organized structures constitutive of life." It seems to me that even if we did not know about the fine-tuning evidence, our experience of creating things, and observing the properties of life, should make us suspect that our cosmos must be fine-tuned to allow for the evolution of such properties. Assume a further principle that what requires most ability also best demonstrates the existence of that ability. Then, given that the production of complex biological organisms requires far more fine-tuning than the existence of something like rocks, biological organisms better demonstrate the fine-tunedness of the cosmos. While these conclusions should also be acceptable to the naturalist who accepts the fine-tuning evidence, they are also of interest theologically. Again, seeing biological organisms as manifestations of the special fine-tuning of the cosmos helps make sense of why a design discourse in biological organisms might ground talk of the Creator better than talk of rocks as dependent on the existence of a ground of being. Given these principles, then theologically, in effect, biological organisms become the metaphorical icons of the temple of nature, which best manifest the inbuilt potential of the entire cosmos, and thus also manifest the wisdom of the Creator.

Wiker and Witt (2006, 242–243) thus argue that we can appreciate the beauty of a biological organism such as a rose better when we realize the extent of fine-tuning that the possibility of that organism depends on. As I have noted (Kojonen 2016b, 373), this also works in the opposite direction: the remarkable fine-tunedness of the cosmos can be better appreciated through observing the marvelous products of that cosmos. The cosmos must be special indeed to allow for the evolution of the kind of complex teleology and the large variety of creatures that we observe. And this feature of the cosmos, the proponent of the design argument can claim, is explained better by a theistic view than by supposing that this feature is due to chance. Thus it seems that biological teleology does retain its revelatory potential even in an evolutionary cosmos, particularly when viewed in the context of a wider natural theology or a theology of nature.

With this basic perspective, I am now in a position to also consider another potential critique. This is that the particular way I have salvaged the biological design argument is of no help to the theist on the street, since my defense also utilizes arcane details of biochemistry and computer simulations, among many other factors. In effect, I have been arguing that the facts of biology, when studied in depth, serve to confirm rather than rebut the basic intuition that the type of complex teleological order we see in biology cannot be produced (or is at least not well explained) without intelligence. This means that it acts as a defeater-defeater against the evolutionary objection to design arguments. However, this does not mean that the theist on the street would not already have reason to see biological design even before knowing about these arguments. Rather, the argument affirms that the theist on the street is essentially correct and that they formed their belief based on good reasons. As Wahlberg points out, we can know based on our own experience (and through the testimony of others) that it is difficult to produce this type of order, and that designing an automated process to produce such order is more difficult still. Moreover, as Nagel (2012, 6) points out, the design intuition (and resistance to Darwinism as an unguided process) is often based on an "untutored reaction of incredulity to the reductionist neo-Darwinian account of the origin and evolution of life. It is prima facie highly implausible that life as we know it is the result of a sequence of physical accidents together with the mechanism of natural selection."[6] The perspective that I have laid out allows for affirming that the untutored reaction of incredulity is supported, rather than undermined, by evolution. This then functions as evidence supporting the reconciliation of the design intuition with evolutionary theory.

Marie George makes a similar point in analyzing Paley's design argument. George thinks we can still defend Paley's (2008 [1802], 75) statement that even if we had no other reasons, the eye "would be alone sufficient to support the conclusion which we draw from it, as to the necessity of an intelligent Creator." George comments:

> In a way this seems a daring statement, but is it really any more than an illustration of the principle evident to the wise and perhaps to all that the order of finality (determinate means to a determinate end) finds an adequate explanation only when one brings in mind? A cause must be proportioned to its effect—one can't get something from nothing, or order from chaos as from an efficient cause.
>
> But there is just the rub: Chance events do seem to give one something for nothing a treasure, a bath, a free lunch without any effort on one's part. Yet, if one contemplates the task of preparing a series of models of organisms, starting with one that was unicellular and ending with man, each of which would be a take-off on the previous one, and all of which would be functional, one realizes that it would take enormous intelligence to do this. Can one really believe that it could just happen by chance? And that is why it seems one could settle the design question simply by looking to the order

> itself, even before one located at what stage the design was introduced and by what means it was realized. So I am inclined to think that design in nature is evident, and that examining necessity allows us to see more clearly, but not more certainly the need for a designer. (George 2002)

This conclusion is wholly opposite to Dennett's (1995a, 133) previously quoted argument, according to which the reducibility of each individual evolutionary event to stochastic chance processes means that the whole series can be reduced to such processes. As George points out, the reducibility of the series to such processes would be an incredibly lucky event, since it would in fact take enormous intelligence to be able to design the kind of environment and series of possible forms that would allow for such a reduction.

We may disagree on whether the evidence from our commonsense reasoning is sufficient to warrant a full certainty of fine-tuning behind evolution. But at least we could say that the theist on the street already has access to reasons that can reasonably motivate the suspicion that there must be more to evolution than just the blind working of chance and necessity. This suspicion is then, in my view, confirmed by the deep study of what sort of tight preconditions must be fulfilled before the evolutionary process can produce complex organs like the eye. Knowledge of these factors can then function as a defeater-defeater against the use of evolution as a defeater against design beliefs that are based on biology. But the fact that the theist on the street also has access to some reasons for this belief personally does mean that the force of biological design as a supporting reason for belief is not wholly dependent on prior knowledge of the fine-tuning case, but can rather serve at least as a powerful clue about fine-tuning.

Analogies of Indirect Design

In Chap. 3, I have utilized thought experiments to better respond to objections against the design argument. As it turns out, plenty of different thought experiments can also be used to argue for the conclusion that the products of a natural process might give us evidence of the Creator of that process, and analyze the implications further.

According to Ratzsch (2001), design detection typically works by first identifying artifactuality through identifying *counterflow*, properties that are contrary to what would be expected based on natural processes.

However, when analyzing artifactual objects, features like complex teleology then function as secondary markers identifying intentional production, as opposed to what can also be accidental artifactual products, such as footprints. Here Ratzsch identifies both *mind correlative* order, which suggests design, and *mind affinity*, which almost inescapably suggests mind. He then argues that while design is typically detected by first observing counterflow, these secondary marks of design can actually provide grounds for design beliefs even in the absence of counterflow. As an example, Ratzsch asks us to suppose that when the first photographs of the moon were taken, we found the text "John 3:16" written on the surface of the moon in craters. Ratzsch (2001, 63) argues that here, "*even if there were a natural explanation for each individual crater*, the craters would constitute an ensemble of a sort for which the conjunction of the individual crater explanations would seem inadequate as an explanation." He further argues that this would be the case even if we could trace the chain of natural explanations all the way back to the Big Bang.

As Ratzsch's example makes clear, we can have powerful evidence of design even in the absence of "counterflow." Even if we were able to explain the physics of the meteor strikes perfectly, and were able to trace the chain of physical causes all the way back to the Big Bang, then this still would not eliminate the evidence of design that the pattern gives us. Rather, the problem of explaining the pattern would merely be pushed back by the physical explanation. In this case, it seems that natural explanations simply do not explain the intelligibility of the pattern, even though they explain each individual crater. According to Ratzsch (2001, 59), in such situations, "if something would constitute evidence of design in the context of some presumed gap in nature, then it will also constitute evidence of design even if the gap in question gets closed naturally."

As I have noted (Kojonen 2016b, 369), evolutionary explanations have been claimed to differ from the kind of physics-based explanations dealt with in Ratzsch's example of John 3:16 on the moon. In the case of this example, tracing the physical origins of the asteroids all the way back to the Big Bang would nevertheless leave the semantic content of the pattern unexplained. But as I have argued in Chap. 4, evolutionary explanations are traditionally understood to explain the same teleological features of biology that intentional design is also supposed to explain. However, the case is claimed to be different in biology. The process of evolution, in this understanding, is not just a carrier of patterns already embedded in nature, but brings about these patterns spontaneously in a way that is highly underdetermined by the laws of nature.

However, in contrast to this perspective, I have argued in Chap. 4 that it is plausible that Darwinian evolution also pushes back the problem of the origin of teleology to the level of the wider teleology of nature, at least to a large extent. If this is correct, then Ratzsch's analogy applies directly to the case of biological evolution. Suppose for the sake of argument that this is correct, and that biological teleology is evidence of design in the first place, as our intuitions seem to indicate. In that case, according to the principles described by Ratzsch, then biological teleology still continues to be evidence of design even after the advent of evolutionary theory.

As I have noted (Kojonen 2016b, 368–369), another analogy that Ratzsch uses is the example of a fully automated factory producing VCRs (video cassette recorders). In such an automated factory it would be possible to explain the physical features of the VCRs as the result of processes within the factory, without any need to refer to interventions by any designer. However, as Ratzsch (2001, 130–131) points out, "we'd still feel that something was missing—that there was something about the factory itself, perhaps implicit in the 'givens' that demanded special explanation." The order of the VCRs themselves, not just the order of the factory, would still call out for explanation in terms of design. But that aside, the designedness of the factory itself is also relevant for explaining the end products: it is "not a simple causal irrelevance, given that without that factory's designedness the [products] would fail both to exist and to have key physical properties they in fact have" (Ratzsch 2001, 130–131). In fact, it seems that the designedness of the factory is inherited by the VCRs, and we would be correct to identify the VCRs as evidencing design. Ratzsch terms this "design inheritedness." As long as it is correct to state that the factory is designed, then it is also correct to state that the VCRs as its products are designed. Moreover, the designedness of the VCRs derives from the designedness of the factory as a whole, and manifests that designedness.

There are vast differences between the environment of biological evolution and the automated factory. However, the general principles of the analogy do seem to be transferable. First, the properties of the factory are relevant for explaining the VCRs, and similarly the environment in which that evolution occurs is also relevant for explaining the properties of biological organisms. If design is a crucial part of this environment, then design is also crucial for a full explanation of the products. The rationality of the factory (and hence the design behind this rationality) is visible in the automatically produced VCRs because nothing within the factory really answers the crucial question of how such order can possibly exist. Similarly, insofar as biological evolution moves the problem of the origin of biological form back to the laws of nature, then nothing within nature itself yet provides a full explanation for how such order can exist. And as argued in the previous section, biological organisms can also visibly manifest the fine-tuning of the cosmos. Still further, it is also correct that if we (in a theology of nature) suppose that the laws that make evolution possible are designed, then this is clearly not a causal irrelevance for the production of biological organisms, just as the designedness of the factory is not a causal irrelevance.

These principles seem to be defensible in many different accounts of evolution. However, the extent to which the analogy of the factory fits the case of evolution does depend on how contingent we believe the process of evolution to be. As noted in Chap. 4, Gould emphasized the extreme contingency of the process, and in such a view the production of life is quite different from the automated and deterministic process which produces VCRs in Ratzsch's automated factory. In such a view, it does remain true that the designedness of the laws of nature is not a causal irrelevance for biology. However, it is also not a causal irrelevance for anything else that happens, whether we are talking about the "course of the wind" which Darwin (1958 [1887], 87–88) mentioned, the shape of the continents, or bank robberies.[7] The potential problem, then, is that God does remain a necessary condition of the emergence of biological complexity, but there is again no special link between God and biological design which would make biological organisms especially expressive of design.

In Chap. 4, I pointed out that this objection was also present in Darwin's (1958 [1887], 426–428) response to Asa Gray's defense of the design argument. In Darwin's analogy of the architect building a house from a random assembly of stones, the shapes of the stones and the laws of the cosmos are certainly necessary conditions for the building of the house. However, if analyzed in terms of a contrastive explanation ("Why did this house get built, rather than not?"), these background factors have very little explanatory power. The architect's actions form the central explanation. Similarly, Darwin argued that while the laws of nature are certainly relevant for making evolution possible, it is nevertheless the historically contingent operation of natural selection which is responsible for how the process of evolution unfolds. The laws of nature are no more determinative of the outcome than in the case of the house built by the architect. Thus we might argue that while God, if He exists, may be a necessary condition of evolution, divine purposes are nevertheless relevant for explaining evolution (Johnson 2015, chapter 7).

In Chap. 4, I argued that this response fails particularly if we adopt an understanding of evolution in which the explanatory role of selection is not quite so all-encompassing, and is further weakened the more we adopt an understanding of "the laws of form" constraining evolution. Even from a Gouldian view, however, in which the course of evolution is indeed extremely contingent, it does still seem that the Creator is not just any necessary condition of evolution, but rather the creator of all the necessary conditions. The distinction between a condition and a cause is sometimes difficult (Schaffer 2016). Nevertheless, it is clear that the possibility of evolution still demands a far higher amount of fine-tuning than something like the creation of a rock, and evolution is still more closely determined by the laws and preconditions of nature than contingent events like bank robberies, at least assuming that bank robberies are caused by genuinely free human agents. In this way, following the arguments of the previous section, I think it is justified to claim that even biological organisms that have evolved through a highly contingent process of evolution would still manifest the fine-tunedness of the cosmos in a special way. However, supposing a structuralist view of evolution, inspired by data like the study of protein evolution and convergence, it seems more plausible to think that evolution is highly directional, and merely discovers forms that are made possible by the laws of nature. This means that Ratzsch's principle of "design inheritedness," which is crucial in the case of the automated factory, applies directly to biological evolution. This makes it even easier to argue that the order of the cosmos is a primary factor driving the evolution of different forms, and organisms thus reflect the designedness of the entire cosmos.

As I have discussed in Chap. 4, an alternative to the Gouldian view is provided by Conway Morris' view based on the prevalence of convergence (Conway Morris and Gould 1998). However, even in this more teleological understanding of evolution, much of the specifics of biological evolution is still left to the working of contingent events. This means that we still need to answer a further question even in such a view of evolution: Can the products of a process incorporating both design and chance elements still manifest design, particularly given that it may be difficult for us to distinguish which elements of the final product are intentional, and which are not? Wahlberg (2012, 181) argues that they can:

> To say that complex biological organisms have been created by God need not mean, as we have seen, that God has 'designed' particular species of organisms. However, since creation is an intentional act, God must have intended something with respect to complex organisms; otherwise it would be nonsense to say that they are God's creatures. Minimally, then, God must have intended that complex organisms in general exist [...] it must also be true that God's intention to bring about such organisms was causally responsible for their emergence.

Previously, I argued that it is sufficient for biological design arguments if design is an answer to the general question of how biological teleology can exist in general, and it is no problem if evolutionary processes are responsible for the details. This still gives God as designer a significant causal responsibility for biological teleology, and in my view suffices to show that perceiving design in biology is not an error in the light of evolution.

Wahlberg (2012, 70) provides the interesting example of computer-generated fugues in order to argue that the products of a design process incorporating random elements can still evidence design.[8] The program writes a piece of music by building on a theme given by the user. The process incorporates random elements: if given the same theme twice, two different fugues will result. Hence, this is an example of unpredictable evolution used in a design process, and neither the user nor the programmer will be able to predict the result (since the programmer does not know beforehand what theme the user will put into the program). Nevertheless, not all properties of the fugue are due to chance:

> It is clear that the programmer is responsible for the fact that what the computer spits out are structures that move within the Western tonal system, have determinate *keys*, are *fugues* (i.e., instantiate a certain form), exemplify certain specific artistic *styles*, etc. These are all properties that the programmer has intended and determined. It therefore seems fair to say that the computer-produced fugues are expressive of the programmer's intent and intelligence in virtue of instantiating certain general (and essential) properties that require a good deal of intelligence to bring about (it is not easy to write a fugue, and neither is it easy to write a program that writes fugues). (Wahlberg 2012, 70–71)

As Wahlberg notes, a person from the eighteenth century would likely erroneously ascribe the fugue generated by the program to design down to its smallest details. However, they would still be correct in assuming that the sounds they hear are expressive of intelligence and intent, since these are so strongly required in order to produce the fugue. Applying this analogy to biological evolution, it is correct that proponents of design have often assumed that biological organisms are designed down to their smallest details. However, applying the results of this analogy, it "could be the case that biological structures are not designed down to the least detail, but that they nevertheless are expressive of intent and intelligence in virtue of being created (in a similar sense of 'created' in which it can be said that the computer-fugues are created)" (Wahlberg 2012, 71). Wahlberg further argues that, in fact, the creation of a program that writes fugues is more difficult and demands much deeper utilization of the principles of music than simply writing a fugue directly. Because of this, one could argue that the products of the fugue-writing program are possibly even more expressive of intelligence than fugues written directly, even though the details are left to chance. Similarly, the products of evolution can also be highly expressive of the divine intelligence that is responsible for evolution, even if the details of evolution are left to chance.

It seems to me that this analogy provides a strong argument for the conclusion that the products of a somewhat random (perhaps highly random) evolutionary process can nevertheless be expressive of purpose. Besides, as Wahlberg (2012, 180) notes, rejecting this conclusion would require assuming that only structures whose every property has been intentionally designed can be expressive of intelligence. But this would be difficult to argue, given that chance elements and unforeseen constraints typically influence the structure of man-made objects, and not every property of such objects is intentional either. Thus, man-made objects typically include what Gould and Lewontin (1979) famously called "spandrels," which are the unintended by-products of the construction process or structure.

Wahlberg also uses the example of architecture, though in a way diametrically opposed to Darwin's example. He argues, rather, that architecture shows how objects created through secondary causes can nevertheless reflect the mental properties of their creator:

> Imagine an architect who has a plan for a castle. She instructs a number of different people to do various things. She tells one person to erect a wall

> here, another to build a door there, etc. Nobody who works on the building knows the plan. Nevertheless, when the building is finished, the architect's plan for the castle as a whole—her idea—is clearly manifested in the overall structure of the building. This is the case despite the fact that she never caused any modifications in the building directly. (Wahlberg 2012, 178)

Wahlberg's example shows that the use of indirect causes does not need to diminish in any way the designer's responsibility for the end result. It seems to me that this analogy can also be modified to demonstrate the compatibility of design with contingent processes. Suppose that the directions of the architect do not fully determine the shape of the castle, but that the architect gives builders significant leeway in choosing among different possibilities for the structure, according to their wishes. For example, the builders could customize the width of the castle's moat, the location of the throne room, and so on. We can also suppose that the builders do not need to implement all of the components planned by the architect, as she has intentionally planned for more possibilities than can be implemented. In this case, it still seems that there is such a strong dependency between the construction process and the original blueprints that we could say that the castle nevertheless reflects the architect's vision, and her plan for construction is a central and indispensable element in explaining the structure. Nevertheless, in this case we could say that the outcome does not reflect only the architect's vision, but also the vision of the builders. This would be analogous to an understanding of evolution in which the process has both directionality and contingency.

So far I have considered mostly examples which support the compatibility of design and natural processes. However, to test the developed principles, it is also important to consider cases where the end result of a process combining design and other processes fails to manifest the designer's existence. Could we construct such an analogy?

To begin, it seems that we can at least extend the analogy of the architect in this direction. We could allow the builders to modify the structure further and add their own plans, instead of relying merely on the original blueprints of the architect. For example, even supposing that the builders are fairly tightly dependent on the blueprints, and do not possess great constructive ability of their own, perhaps modifying a section of the courtyard into a prison instead of a playground would be within their ability. They would need to just make the walls much higher, and remove the possibility of opening up the gate from the inside. While minor modifications would still let the final structure reflect the architect's plan, there may come a point where the modifications and additions made by the builders make the architect's vision barely distinguishable. For instance, while changing a house into a prison would require seemingly minor modifications in structure, these would nevertheless have drastic implications for the purpose of the structure. In that case we could still say that the architect's vision is a core explanatory component of the castle, but arguably this would diminish the architect's responsibility.

We might be tempted to say that this version of the architect analogy comes close to Gouldian, highly contingent evolution, as opposed to the less contingent understanding of evolution presented by Conway Morris. However, this is not clear as there are still many differences between God and the architect. The architect works through builders who exist regardless of the architect's actions, and who have independent agency and planning capability. In contrast, all evolutionary factors are created by God and ultimately depend on God, nor do evolutionary processes have conscious agency. To see the importance this difference makes, we must again modify the original example. If the builders in the example were robots created by the architect rather than autonomous actors, and if the architect was capable of fully predicting what each robot does, then it would be strange to think that any action by the robot-builders could make the architect less responsible for the final structure of the castle. Similarly, with evolution, if all historical contingent factors are also within God's providence (even though animals are not comparable to robots, but have some autonomy), then the existence of such factors would be part of God's design plan. Still, the dependency of the organisms on the original design plan is clearer and simpler in the structuralist understanding of evolution.

Continuing the analysis of this example would lead to deep questions on the nature of freedom and responsibility, and in what sense this could apply not only to humans, but also to biological organisms as agents whose actions unconsciously affect the evolution of their own species. The issue of responsibility for evil is also raised by the next example, the evolution of technology.

Another analogy comes from human technology, following Darwin's analogy of architecture. It seems that the laws of physics are indeed necessary conditions for human technology, but not the central difference-making factors for explaining what kind of technology we specifically develop and use. In Chap. 4, I have already referred to Dennett's (1995b) critique of Goodwin's structuralism, according to which the idea of laws of form in biology is just as ridiculous as the idea of laws of automotive engineering. If evolution has merely discovered biological forms, then could we not also say that engineers have merely discovered the forms of televisions and cars, rather than actually designing them? This, then, seems like a potential *reductio ad absurdum*-style argument against my position. Suppose that we think that because of the strong dependency of biological evolution on laws of form and fine-tuning, evolution has not really designed the various biological organisms from scratch, but has simply discovered forms made available by the way nature is ordered. Because of this, then, we assert that God as the creator of the cosmos is ultimately responsible for biological teleology, and his existence is manifest in the existence of biological teleology, even if the actual production of the organisms is done through an evolutionary process that combines and recombines the possibilities that God has made available. Could we not then say the same or at least a similar thing regarding human technology? And would this not be clearly false, because we know that humans are actually responsible for creating these artifacts? Surely we would not even want to credit God as the creator of all human technology, such as instruments of torture.

It does seem to me that a strong argument can be made for the thesis that the development of technology also requires special fine-tuning (Collins 2018). The relative simplicity of the laws of nature and the applicability of mathematics to nature are important enablers of scientific discovery. Wigner's (1960) classic statement of the problem is that "the miracle of the appropriateness of the language of mathematics for the formulation of the laws of physics is a wonderful gift which we neither understand nor deserve."[9] In addition to the dependency of science and technology on the possibility of mathematics, many authors have argued that the cosmos also seems to have other features that facilitate the development of science. For example, it has been possible to gradually advance in our understanding of the cosmos from simpler to more complex theories, when it might as well have been constructed in a way that requires the more complex theories to be understood first. Thus we do not need to begin with theories of general relativity and quantum mechanics, but can first develop a commonsense account of physics, then move on to Newtonian physics, and so on. Wiker and Witt (2006, 245) thus argue that, despite the arduousness of scientific discovery, "the universe is crafted to condescend to our capacities as a teacher to a student and to draw us patiently upward; and the superabundance of intelligibility is a sign that it was made by a mind that far exceeds the merely human."

If the development of technology depends on extensive fine-tuning, then it does seem that the products of human technology could also manifest not only human wisdom, but also the fine-tunedness of the cosmos, and by extension the wisdom of the Creator. Nevertheless, there are a number of disanalogies with biological evolution which seem to make the manifestation of the Creator's wisdom weaker. For instance, biology seems to require more fine-tuning than the development of technology. Human designers have the ability to bypass the need for functional intermediate forms on the way to a new application that allows them to build forms that would be inaccessible to a process of natural selection. Moreover, humans as free agents have the possibility of inventively combining technologies and modifying objects for their own ends, and their freedom also diminishes the Creator's responsibility for the particular devices that humans create, including both beneficial and evil applications.

Of course, in the theistic view, all human behavior and interaction, including the possibility of consciousness and free will, will also be dependent on the creation and ordering of the cosmos. The position of theological determinism—the idea that God determines all events—has also had its proponents in the Abrahamic theological traditions. In such a view, it does seem plausible that all responsibility for human action will also ultimately be laid on God, including evil actions. Indeed, this is one of the core theological problems of divine determinism. But most theological positions do allow for the existence of at least some human free will, so that humans are not to be considered simply as extensions of God's will, or as indirect causes through which God acts (List 2019). This at least seemingly creates a large difference between the evolution of technology and the evolution of biological teleology—though, as we will see, many would want to also absolve God of the responsibility for evolution by referring to the freedom of the process.

5.3 Dealing with the Problem of Natural Evil

As noted, the problem of natural evil is one of the traditional main arguments against any natural theology. Christopher Southgate, for example, persuasively argues that a natural theology should take into account both the good and apparently evil aspects in order to be true to reality, and not dwell merely on "the beauties of creation, on verdant meadows and coruscating sunsets" (Southgate 2018, 103). Many different problems have been proposed. For instance, it has been argued that something about evolution itself already constitutes evidence against theism, because the evolutionary struggle seems disconsonant with God's loving nature (Draper 2012; contra Peels 2018). In this sense, then, it might be argued that while the apparent teleology and beauty of the biological world provides evidence that fits well with theistic belief, the cruelty and ugliness of nature provides compelling counterevidence. Many also see the evolutionary process itself as adding to this cruelty. Others, however, argue that evolution actually makes theism more defensible by showing that the apparently badly designed properties of nature are actually the products of free natural processes that God has merely put in motion but does not control. According to Ayala (2007, 160), for example, the problem of evil makes any claim that the present features of the creation are divinely designed theologically dangerous: "attributing these to specific agency by the Creator amounts to blasphemy."[10]

Understanding the Problem

Perhaps surprisingly, both the idea of natural suffering as counterevidence for theism and the idea of evolution as a solution to the problem can be found in Darwin's thought. In his autobiography (1958 [1887], 52), he argued that the omnipotence and perfect goodness of God are difficult to harmonize with "the sufferings of millions of lower animals throughout almost endless time." Nevertheless, in an earlier letter to Asa Gray, Darwin also provided one possible answer, in a text that also provides one of Darwin's most extensive reflections on the relationship between design and evolution:

> With respect to the theological view of the question; this is always painful to me.—I am bewildered.—I had no intention to write atheistically. But I own that I cannot see, as plainly as others do, & as I shd wish to do, evidence of design & beneficence on all sides of us. There seems to me too much misery in the world. I cannot persuade myself that a beneficent & omnipotent God would have designedly created the Ichneumonidae with the express intention of their feeding within the living bodies of caterpillars, or that a cat should play with mice. Not believing this, I see no necessity in the belief that the eye was expressly designed. On the other hand I cannot anyhow be contented to view this wonderful universe & especially the nature of man, &

> to conclude that everything is the result of brute force. I am inclined to look at everything as resulting from designed laws, with the details, whether good or bad, left to the working out of what we may call chance. Not that this notion at all satisfies me. I feel most deeply that the whole subject is too profound for the human intellect. A dog might as well speculate on the mind of Newton.—Let each man hope & believe what he can. (Darwin 1860)

Thus, Darwin presented a theological argument against the idea that God specifically designed individual organisms. This would, says Darwin, mean that God is also responsible for designing repulsive creaturely behaviour, such as parasitism, and this seems (for Darwin) to be in tension with divine goodness. Nevertheless, Darwin also does not think it tenable to believe that "this wonderful universe" is the result of brute force, rather than design. His tentative solution is to suggest that the Creator has designed the laws of the cosmos, including those driving evolution, but has left the details up to chance.

The severity of the problem depends partly on our estimates of nature. As McGrath (2016, 74) points out, "We feel disturbed by the presence of suffering in the world. Yet the question of the reliability of these intuitive human judgments about nature cannot be overlooked. It is highly questionable whether any aspect of the non-human natural world can be described as 'evil.' It is certainly 'natural'; yet the judgment that it is 'evil' demands a framework of evaluation which ultimately rests on higher-order beliefs and assumptions going beyond nature itself." For example, regarding Darwin's specific example of the *Ichneumonidiae* wasps, it can be asked whether this is really that serious as a problem, since the capacity of insects to suffer is actually very limited, even if this is not taken to the extreme of denying the existence of morally relevant suffering in many animal species (Miller 2020). Garvey (2019), for example, thus argues that the extent of the problem of natural evil has been vastly overstated ever since the Reformation period, and that the good in nature vastly outweighs the bad. Moreover, against the idea that evolution is "red in tooth and claw", contemporary accounts provide some balance by pointing to the importance of cooperation in evolution (Nowak and Coakley 2013).

Paley famously argued that the capacity of organisms for pleasure and the typically beneficial nature of biological adaptations provides powerful evidence for the goodness of the Creator, enough to allow for explaining the remaining evils as part of the morally necessary tribulations of this life. In later discussion, Paley's "happy world" has often been criticized as not giving sufficient weight to the prevalence of suffering in nature, particularly predation (Fergusson 2018, 201–211). But in Paley's defense, he does spend time in his *Natural Theology* discussing predation, death, and disease, and indeed the book itself was written while Paley was himself suffering from illness (Paley 2008 [1802], chapter XXVI). Thus, different estimates of the difficulty of natural evil as a problem cannot be explained simply on the basis that some are unacquainted with the existence of suffering in nature. In the contemporary discussion, aesthetic theodicies which aim to contextualize evils are still relevant (Schneider 2020).

Nevertheless, whatever we think of the overall goodness of nature, and whether we find Darwin's own examples convincing or not, it does seem that something about the appearance of natural evil stands in need of explanation. However, evaluating the extent of the problem and all of the possible responses would require too much space to attempt here. The evaluation of nature is a complex matter, and calls for intellectual humility (Dougherty and McBrayer eds. 2014; McLeish 2014, 102–148).The literature on the problem of natural evil is vast, and many different responses have been made (Murray 2011; Southgate 2008; Dougherty 2014; Stump 2010). Thus, instead of attempting to respond to the entire problem, I will focus here on the relationship of the previously mentioned evolutionary theodicy with my approach.

I have argued that evolution does not remove God's indirect responsibility for the beauty and ingenuity of the teleology we see in biological organisms—so much so that these structures give evidence of divine purpose behind nature. However, does this conclusion not then mean that I should also have to blame God for the apparently malicious features of parasites and dangerous pathogens? In this section, I will argue that insofar as the kind of evolutionary theodicy proposed by Darwin, Ayala, and others works at all, it is indeed also compatible with the salvaged biological design argument. I will begin by evaluating the basic premises of the evolutionary theodicy, and then move on to consider it in relation to my approach.

Related to the broader problem of the evidential effect of the problem of evil, I have argued that the evidence of design does not need to provide us with a proof of God's existence or goodness in order to be valuable. That this evidence merely supports the rationality of religious belief would thus already be an interesting conclusion, particularly as it runs counter to the currently prevalent interpretation of biology. The design argument could then function as part of the "problem of good" counterbalancing an overt focus on evil. As Boethius (475–526) asked in *The Consolation of Philosophy* (part IV), "If God exists, whence comes evil? Yet whence comes good, if He exists not?"[11] Thus understood, the design argument is compatible with both a Swinburnean cumulative case argument and something like McGrath's Christian natural theology as a project of "faith seeking understanding"—or with natural theological approaches from other religious traditions. In all these cases, the broader theological and philosophical worldview will affect the response to the problem of natural evil.

Indeed, due to the seemingly clear existence of at least some less than perfect features of nature, as well as biblical affirmations that the creation is not yet what God ultimately intends, constructing one's understanding merely based on the creation seems theologically problematic. Luther, for instance, stated that "We could never attain to the knowledge of the grace and favor of the Father except through the Lord Christ. [...] outside of whom we see nothing but an angry and terrible Judge" (Luther 1921 [1529], III, 65). As Loikkanen (2015, 181) notes, as in some other questions, Luther may not have been fully consistent here (he often, for example, uses hyperbole to emphasize points that might need to be stated in a more nuanced way). In his *Lectures on Genesis*, Luther (1964, 113) also wrote that "all men naturally understand and come to the conclusion that God is some kind of beneficent divine power, from whom all good things are to be sought and hoped for."[12] But generally, emphasizing the insufficiency of the natural order and the idea of the centrality of Christ for full knowledge of God is shared by all Christians. Both Southgate (2018) and McGrath (2016) also favor interpreting nature's revelation of the Creator in the context of Christ's cross and eschatology. Other theological and religious traditions will likewise interpret the problem of evil in their own context.

I will turn, then, to the more limited question of how the previously discussed evolutionary theodicy relates to this book's argument. The idea is that insofar as God had to (or had good reason to) create through a free evolutionary process, then this helps explain why the features of nature are not optimal, and not fully reflective of the divine character. In contrast, for Ayala and others, the biological design argument is not compatible with this strategy, since it seems to require thinking of the properties of individual organisms as reflective of divine intentionality. Thus the Creator becomes responsible for parasites like the *Ichneumon* wasps, and cats toying with mice. Not all proponents of design arguments have shied away from this seemingly theologically problematic conclusion. Behe, for example, bites the bullet and argues that the malarial parasite displays clear signs of design: "Here's something to ponder long and hard: Malaria was intentionally designed. The molecular machinery with which the parasite invades red blood cells is an exquisitely purposeful arrangement of parts" (Behe 2007, 237).

Behe goes on to argue that the design argument's conclusion is not made invalid by the theological questions this conclusion raises. Because the design argument itself, as formulated by the ID movement, does not rely on assumptions about the goodness of the designer, it is, Behe claims, also compatible with the hypothesis that the designer is indeed a moral monster. Thus Behe (2007, 238) claims that the design argument, by itself, does not determine whether the designer is a "dope, a demon or a deity." Rather, he leaves it to theologians to ponder how to reconcile malaria with the Creator's goodness. I agree that based on our common human experience, we can indeed recognize objects as designed, even if they are not perfect by moral or engineering standards. However, I have argued elsewhere that the distinction does not allow ID to bypass discussion of natural evil, because in practice the credibility of ID's design arguments is influenced by the relative credibility of theism and naturalism (Kojonen 2016a, 156–162). For this reason, it would be important for proponents of ID to have a model that allows one to explain apparently malicious design.

Despite this critique of ID, I nevertheless agree with ID proponents that evolutionary theodicy is just one among many different responses to the problem of evil. Thus ID proponents can appeal to these other reasons, even if not to evolutionary theodicy. Moreover, their approach also permits them to argue that God allows for chance processes to explain some features of biological organisms. Dembski, for instance, only claims that his eliminative design inference can eliminate chance and necessity as sufficient explanations for the patterns of life; it does not eliminate them from having any role at all.[13] Even Paley (2008 [1802], chapter V) allowed for chance in explaining features like warts and moles. However, as noted, in the case of malarial parasites Behe himself argues for the necessity of design as a cause, claiming, as noted, that the production of such an exquisite design is beyond the capacities of unguided natural processes. This clearly rules out the possibility of an evolutionary or chance-based theodicy as accounting for this instance of natural evil.

It seems, then, that proponents of ID must explain the origin of the malarial parasite (and other similarly orderly pathogens) as the intentional product of design (whether divine or demonic), whereas theistic evolutionists can present a model in which no designer is directly responsible for the malarial parasite (assuming that this is not invalidated by the fact that God made the evolutionary process possible). In that case, proponents of ID are potentially taking on a major disadvantage in dealing with the problem of natural evil. And perhaps, then, a similar argument could also be raised against any salvaging of the biological design argument. If we want God to receive the glory for biological teleology, then should God not also receive the blame for malaria, parasitism, pathogens, and bad design generally? How could divine design only be a proper conjunctive explanation for the beneficent teleology of life, and not be responsible for malaria? Would it not be better to give up on trying to salvage the biological design argument rather than believe that God has intentionally designed such things?

To answer these questions, there is a need to further analyze the notion of causal and moral responsibility. One possibility would be to argue that the evolutionary theodicy does not actually work in responding to the problem of evil, and in that case adopting the biological design argument would create no particular disadvantage. But another possibility would be to claim that the version of the biological design argument developed here can actually adopt the evolutionary theodicy just as well as any other theistic evolutionist approach, since it does not require that God is solely responsible for the order of biological life. In the remainder of this section, I will explore these questions by analyzing two foundational assumptions of evolutionary theodicies. First (1), the credibility of evolutionary theodicies depends on the supposition that a process-like method of creation produces some valuable benefits in comparison to a miraculous creation, and further that evolution was the best (or only) way to use a process to create living species. Second (2), defenders of any evolutionary theodicy must show that such a process would at least mitigate divine responsibility for the products of that process. To do that, we would have to suppose that the process can somehow incorporate real contingency and freedom, with the results at least somewhat undetermined by God, even despite the doctrine of creation out of nothing (Russell 2013). I will now tackle each facet of this question in turn, and in the process will also discuss how the salvaged biological design argument can incorporate an evolutionary theodicy.

Evaluating Evolutionary Theodicies

Is it credible, then, that evolution was the only way for God to create life? Or if it was not, could we at least present a good theological rationale for evolution? For many thinkers, it has seemed unlikely that God would have used an evolutionary process that would lead to apparently wasteful or badly designed outcomes. Rather, it has seemed more likely that God would have used another type of process, or would have created all living species miraculously, without the need for an evolutionary process. Both arguments for divine freedom and arguments regarding the suffering inherent in an evolutionary process are crucial here (Draper 2012). Nevertheless, others contend that, at least in retrospect, creation through an evolutionary process also has some unique value.

The assumption of divine freedom in creation seems secure for traditional theists. In the traditional understanding, God's act of creation is free, meaning that God could have created a wholly different type of world or could have abstained from the act of creation completely. What possible theological reason could we then have for assuming that evolution was the only way that God could have created living species (Russell 2013)? Certainly most Christians who have lived did not think that God created life through an evolutionary process, and thus it seems difficult to think that evolution is a necessary component of the Christian doctrine of creation. After its discovery, evolution does need to be included in discussions on creation; and it does fit previous discussions about creating through natural causes, beginning with Augustine's account of rational seeds (McMullin 2009), but the theory could not have been derived from theological doctrines alone without recourse to empirical evidence.

In response to the argument from divine freedom, it has been argued that creation out of nothing may also be subject to logical and metaphysical constraints, and perhaps these constraints (unknown to us) make it so that an evolutionary process involving suffering truly is the only kind of natural process available, though our grasp of such necessities that seemingly bind even God is faulty (Southgate 2008). For example, perhaps the existence of a law of natural selection is a necessary consequence of the existence of mortal creatures with finite resources. Perhaps acknowledgment of our human limitations could at least allow us to state with McGrath (2009, 204) that "We are not in a position to demonstrate that there is, or could be, a superior way of constructing this world, or bringing life in general, and humanity in particular, into being. We may complain about the existence of pain and suffering. But that hardly amounts to a disconfirmation of the goodness of the world."

However, it seems to me that because the idea of evolution as the only way is at least in tension with the idea that God's creative act was free, the theodicist should indeed try to argue that Darwinian evolution was a good way of creating, or even the best way of creating, the species, rather than the only way.

Here the evaluation of the amount of suffering in the evolutionary process will affect how the morality and beauty of the evolutionary process is evaluated, as will the broader theological framework. As noted, Christian eschatology leads to the idea that the present world is hardly perfect, and many would thus see the suffering in evolution as reflecting our current state in the process of creation, rather than God's ultimate purpose for the creation. Peters and Hewlett (2003, 173) thus argue that natural theologies which do not take the Incarnation and the eschatological perspective into account are in danger of justifying present-day evils: "the designer imagined can be nothing more than the designer of violence, suffering, death and enormous waste. Nature does not all by itself [...] offer the promise of renewal." They further argue that "[b]ecause God sides with the victims of suffering, our constructive proposal for a theistic interpretation of evolution cannot invoke a divine blessing on the survival of the fittest, either in ontology or ethics. [...] The struggle for survival is a mark of fallen creation" (Peters and Hewlett 2003, 172). Peters and Hewlett emphasize the need to view the imperfections of the present creation in light of the coming redemption of creation. This means that features of the present world like death and suffering cannot be seen as indicative of God's ultimate purposes and nature (similarly Dougherty 2014; Schneider 2020, 219–269).

Nevertheless, to take this view to the extreme of affirming survival of the fittest as evil would be taking it quite far. In a theistic evolutionist view, the survival of the fittest is a central part of the present order of nature, and a central part of the mechanism that God has used to create the variety of living beings. If our theological view terms such a basic feature of reality as morally evil, or as part of the fallenness of creation, would this not mean that the Fall predated even the creation of life? The tradition does speak of the fall of angels, which may well also be applicable to the problem of natural evil (Peckham 2018), and some have argued that this would allow us to understand natural evil as the result of a cosmic struggle between good and evil (Peckham 2018; Hart 2005). These accounts are meant to avoid the problematic idea that these features of nature are directly intended by God, and thus to avoid justifying evil (a problem highlighted by the critique of theodicies, Snellman 2019). Nevertheless, to say that this affects such a basic feature of the created order would be to take this quite far. Would it not be desirable to say that there is something of value in the evolutionary process (Peels 2018)?

Moreover, as things currently stand, it is difficult to imagine what a world without survival of the fittest would look like, even if the world was created without evolution. As Peels (2018, 559) points out, "A process that does not favour the well-adapted over those that are not well-adapted will, in the long run, lead to species that are radically unadapted in the sense that their chances of reproducing in the relevant environment are either negligible or zero."[14] In an orderly world with finite, mutable, and mortal creatures who require resources to survive and reproduce (with finite resources available), it is arguably inevitable that some natural selection will occur. As Cunningham (2010, 76–77) also notes, creatures must seek to protect themselves and their own progeny, or they would become extinct. It would require constant miracles for creatures that are less adapted to their environments to be able to do this as well as those who are adapted. Though the Christian message requires that the human animal should ultimately transcend selfishness, perhaps, then, this should not lead to the condemnation of self-preservation as such. But if so, then natural evils are related to phenomena that were already known prior to the advent of evolutionary theory, such as predation, parasitism, disease, hunger, and death. Perhaps it could be argued that there is no reason for God to create mortal creatures capable of suffering in the first place, but that is an argument that is separate from evolution as such.

But what kind of positive reasons could God have for choosing to create through an evolutionary process, even if we suppose that natural selection will inevitably happen given mortality? Here theologians and philosophers of religion have attempted to argue for a variety of reasons, such as (1) the proposed intrinsic goodness of an evolutionary process, (2) that the evolutionary process helps create an epistemic distance between humanity and God, which is argued to be necessary for creaturely autonomy, and (3) that the autonomy of creatures is argued to be better ensured by an evolutionary process, and the regularity of natural processes in general.

All of these reasons are controversial. For instance, Wahlberg (2015) argues that, contrary to the only way argument, creating through evolution would not be in any way necessary for at least most of the goods that are argued to result from evolution, such as creaturely autonomy. To argue this, he asks us to imagine a twin Earth, identical in all respects to our cosmos, but created out of nothing 200 million years ago, rather than now. The copy is fully identical to the present Earth, and God relates to it in the same way as he relates to our cosmos. The only difference is that the twin Earth lacks our long evolutionary history. However, it seems far from obvious that this difference would matter for the autonomy of current creatures, given that the current properties of the cosmos and these creatures are the same. Thus, since it seems that the same results can be gained even without using an evolutionary process, it seems that an evolutionary theodicy should find some value in the process itself, rather than merely in the consequences of the process for present-day creation.[15]

One way of arguing for this is to say that somehow creation through a process reveals the divine wisdom better, or otherwise fits with divine nature better. There is strong support for divine action through secondary causes in the Christian theological tradition, but the present preference for a process-like creation is also heavily influenced by Enlightenment-era metaphors of the cosmos as a divinely engineered machine. Murray (2011, 144–145), for example, quotes Leibniz' argument against the interventionism of Newton:

> Sir Isaac Newton, and his followers, also have a very odd opinion concerning the work of God. Acceding to their doctrine, God Almighty wants to wind up his watch from time to time: otherwise it would cease to move. He had not, it seems, sufficient foresight to make it a perpetual motion, Nay, the machine of God's making, is so imperfect, according to these gentlemen; that he is obliged to clean it now and then by an extraordinary concourse and even to mend it, as a clockmaker mends his work; who must consequently be so much more the unskillful a workman, as he is more often obliged to mend his work and set it right. According to my opinion, the same force and vigor remains always in the world, and only passes from one part of matter to another, agreeably to the laws of nature, and the beautiful pre-established order. And I hold, that when God works miracles, he does not do it in order to supply the wants of nature, but those of grace. Whoever thinks otherwise must needs have a very mean notion of the wisdom and power of God. (Leibniz 1956, 11–12)[16]

The Leibnizian line of argument is that a process-like creation better glorifies God, and the necessity of interventions by a Creator in natural processes would reveal the Creator to be imperfect, an "unskillful workman." This would fit well with the possibility that the Creator acted through an evolutionary process. Indeed, Leibniz has sometimes been seen as a forerunner of evolutionism.

However, this argument also does not seem conclusive. It seems possible to ask whether the appeal of this argument is based on the selected metaphor of machinery, and if another conclusion might seem more credible if we adopt a different metaphor. Criticizing the idea that God should act in natural history only through secondary causes, Dembski (2002, 328) asks us to imagine the cosmos as a musical instrument rather than a watch. Unlike watches, musical instruments are made to be interacted with, and there is nothing strange about playing a piano one has made. Similarly, if we imagine the cosmos as a temple or as a kingdom, there would be nothing strange or imperfect about God's interventions in the world. The Christian tradition speaks of divine action beyond the laws and processes of nature, and on this basis we could argue that the overt reliance on metaphors which deny the possibility of interventions is misguided. Indeed, Leibniz himself also believed in the actuality of miracles in history. I have also argued elsewhere that the possibility of interventions cannot be excluded simply on theological grounds, but should be based on our evaluation of the evidence in each case (Kojonen 2016a, 102–106; Kojonen 2019).

Still, Leibniz' analogy does show that creation through a process that does not require intervention can indeed be beautiful and valuable. Thus, it provides a possible reason that could help explain why God could have used secondary causes in his creation—provided that the secondary causes used (such as evolution) are not themselves morally repugnant (which depends on our overall evaluation of the problem of suffering in nature). One possibility that would fit the salvaging of the biological design argument well is related to Leibniz' contention that creating through a process arguably reveals God's wisdom and power better than creating directly. Murray (2011, 146) thus argues that "there is something grand, beautiful, and artful about a universe which contains within it everything that is necessary in order for it to bring about the results God intends for it." As noted previously in this chapter, based on our experience it takes much more skill to produce something through creating an automated factory that produces the end result than to produce the end result directly. Thus, though evolution has, for many, given us a way to explain design without a designer, Murray and others would argue that creating through evolution manifests the glory of the Creator in greater depth than would creating directly. Here, then, I can already note that this line of argument seems to actually depend on the possibility of saving the biological design argument, rather than being undermined by it. If it were the case that evolution really removes biology's order as evidence of the Creator, then it would not seem that evolution better manifests the Creator's glory. Rather, in that case we would have to argue instead that evolution is valuable because it makes God more hidden.

Wahlberg's (2015) argument from a "twin Earth" can also be responded to by demonstrating other values that result from the actual long history of the cosmos, rather than merely the appearance of such a history. While complaints about the suffering of billions of creatures over a long evolutionary history are common in the literature, it should be noted that with the increased length of history, we do not only increase the amount of suffering, but also the amount of pleasure. Thus, the long history of the cosmos has enabled the existence of a far greater number and variety of creatures than would have been possible in a briefer history. Surely the long-term coexistence of, say, an advanced human civilization and vast numbers of dinosaurs would have presented challenges. Yet it does seem to be a good thing that dinosaurs have existed, and that they (in their non-bird varieties) continued to exist for nearly 200 million years. If we move from an anthropocentric perspective to a biocentric view, in which the existence of the variety of animals has value independently of humans, then it becomes clear that the long history of the cosmos implied by evolution does indeed add some value (Sterba 2019). And even in a Young Earth view, it seems that God would nevertheless be responsible for creating predators and prey, or responsible for allowing the creation to be fundamentally changed due to human sin (Murray 2011, chapter 3).

It seems, then, that the effects of evolution in making the problem of evil worse have been exaggerated. Nevertheless, evolution might invalidate some theodicies and enable others. The theological explanation of evil that is often thought to be invalidated by evolution is the explanation based on the doctrine of the Fall. Young Earth creationists commonly explain the existence of both human and animal death as the result of the human fall into sin. But this is not easy to harmonize with the existence of death temporally before the human fall, and with death as part of the evolutionary story that explains how humanity came to be.[17] However, even in an evolutionary view, human sin can still explain some natural evil (e.g., natural disasters and droughts as the consequence of human action), and at least some natural evils could similarly be explained as the result of the fall of angels even prior to the existence of man (Peckham 2018), though, as stated, applying this to explain basic features of reality like death seems doubtful.

Therefore, regarding the first assumption of evolutionary theodicies—the assumption that God had good reason to create through evolution—it does seem possible to argue that creation through evolution has some benefits. This assumption, it seems, can actually be defended in a more robust way given a design-based perspective—given the point that a process-like creation might also manifest the Creator's wisdom. However, the theological case for evolution is far from conclusive, and it seems necessary that theology must here simply follow the evidence of how God has created.

I turn now to the second assumption of evolutionary theodicies. Let us suppose that God does have a good reason to create through evolution, or even that evolution was the only way to create. Would this then show that God is not responsible for the products of that process? It seems to me that here the problems are much more severe than in the case of the first assumption, and the assumptions regarding divine causal and moral responsibility are problematic. However, insofar as these assumptions are defensible, then it seems to me that this theodicy is indeed compatible with the biological design argument.

The Freedom of Evolution and the Responsibility of the Creator

There have been several defenses of the value of evolution for fostering the freedom of God's creatures (Southgate 2008; Moritz 2019; Sollereder 2018). For example, Sollereder (2018, 116–117) argues that:

> The freedom of life, seen in the evolutionary process, can result in methods of survival that do not necessarily reflect God's design or purpose, but are coherent from within a perspective of the freedom extended by divine love. Therefore, God did not design the parasitic Ichneumon wasp to lay eggs within a living host, but God's love created an open field of possibilities in which the Ichneumon would have the freedom to develop that survival technique as it pursued its desires.

This, together with other similar approaches, essentially applies the free will defense to natural evil. The free will defense (Plantinga 1977) is the most discussed reason why God might have allowed the existence of human moral evil: because God valued the existence of genuinely free creatures, he chose not to determine the nature of his creation solely by himself. Rather, the free creatures made by God are co-creators with God in the sense that both their good and evil choices will influence what the world will be like. Sollereder refers to the value of the "freedom of life" to explain why God has allowed the evolution of less than morally perfect, even malicious, features in biological organisms. God, as love, allows all of his creatures to be free, and this includes allowing them to act in ways that influence evolution. This means that God cannot allow only those actions which would result in the evolution of cooperation and altruism, but must also allow evolution which results in parasitism and predation, for freedom is not real if God only allows that which fits with his intentions.

As Sollereder (2018, 105) puts it, love "requires making place for the freedom of the other [...] the freedom of creation is the freedom for creatures to act with real causative effect on earth history without their behaviour being fully determined by God." Thus, then, love also requires giving each individual creature freedom to act. Collectively, the free acts of millions of creatures, pursuing their desires, will then result in trends in evolution that might contradict God's desires. Even though many survival strategies cause suffering to other creatures, God cannot intervene without revoking his gift of freedom. However, Sollereder does not assume that biological organisms have free will in the libertarian sense, and prefers instead to simply describe them as agents. In her understanding, since God values agency and freedom so highly, then God must value even the limited agency that biological organisms have, allowing it to realize all possible evolutionary outcomes. This evolutionary theodicy would thus fit well with views emphasizing the fully uncontrolling nature of divine love (Oord 2015).

As I have argued in this chapter, it seems to me that the biological design argument is compatible with Darwin's view that God created living species through fixed laws, while leaving the details up to chance. I have argued that the end products of a designed process can provide evidence for and make manifest the existence of the designer, even if that design process makes substantial use of chance processes, and the end result is not fully determined by the designer. Thus, it seems that the salvaged design argument is compatible with at least a great deal of freedom and contingency in the evolutionary process. Thus, if there really is freedom in the evolutionary process, and/or in the actions of biological creatures, as

Sollereder and others argue, this is compatible with the salvaged biological design argument, as long as something substantial about biological teleology remains to be explained as the result of divine design. As Wahlberg (2012, 185) puts the point, the malarial parasite can be seen as "expressive of divine intent and intelligence by virtue of being capable of displaying rather clever functional behavior. However, the specific kind of functional behavior it displays need not, if I am right, have been intended by God." Southgate (2018, 143–144) similarly concludes that the malarial parasite, though not being expressly intentioned by God, and though its way of life is harmful to other creatures, nevertheless "expresses something of the fecundity and generativity of creation. As such it too, hard and troubling though it is to say, is an aspect of the divine glory."

Evolutionary theodicy, then, seems in principle to fit very well with the approach I have taken for salvaging the biological design argument. As I have formulated it, the argument does not require God to be responsible for all the details of biology in order for his glory and wisdom to be manifest in it. However, as appealing as the possibility of such an evolutionary theodicy is, it raises difficult questions that have not been fully responded to in the literature.

As noted, Sollereder's argument appeals to the freedom of animal agency, rather than libertarian free will as such. Nevertheless, in the case of humans, the plausibility of the free will defense rests largely on the existence of libertarian free will, meaning the capacity of creatures to choose to act otherwise (undetermined by previous factors, though still influenced by them), and to be the source of their own actions (Timpe 2013). If we can affirm libertarian free will, then it becomes intelligible that God would not be responsible for my actions, and that I am primarily responsible for my own actions. For instance, if I choose to rob a bank, then it can be plausibly claimed that God is not responsible—even though the bank robbery could not have happened without God's decision to create the world. However, the reality of such a robustly free will is sometimes contested even in the case of humans, and it would be unheard of to suggest that other animals possess libertarian free will or even moral responsibility in the same sense as humans do (Tomasello 2019; Henrich 2015). Let us affirm that organisms can, through factors like sexual selection and niche construction, also influence the course of their species' evolution. Collectively, the individual actions of organisms will greatly affect the direction that evolution takes, and the importance of these factors is emphasized in the extended evolutionary synthesis. Nevertheless, the ability of each individual organism to do this is limited, nor are these organisms even conscious of evolutionary trends, so that they could make real moral decisions concerning them. Thus, it seems doubtful that their collective effect on evolution is truly free and autonomous enough to absolve the Creator of responsibility, or to show why the Creator would or should not have limited evolutionary outcomes.[18]

To elaborate, in the discussion of moral responsibility, this is typically assumed to have two main conditions: the ability to control one's actions and the ability to know the consequences of one's actions (at least to some degree) (Talbert 2016; Visala 2019). Here we can also distinguish between moral and causal responsibility. For example, if a fire happens due to an electrical fault in my stove, after I put the stove on, I would be causally (partly) responsible for the fire, since the house would not have burnt down if I had not used the stove. However, if I did not have reason to suspect that the stove was faulty, I am not morally responsible (unless we can argue that I should have known, or should have made sure before using the stove). The causal distance between my actions and the consequence also matters for moral responsibility, since it will be difficult for me to predict the consequences of my actions.

In the case of the effect of biological organisms on evolution, the causal responsibility of each biological organism is minute, and arguably their actions are mostly determined by their instincts. This means that the condition of control is unfulfilled. The moral responsibility of each biological organism is clearly non-existent or extremely small, since none of them will be able to even understand the effect their actions will have on the future direction of the evolution of their species. The biological organism's lack of knowledge about the evolutionary effects of their actions is also not due to any mistake on their part, given that they do not even have the cognitive capacity to understand evolution. In contrast, in the traditional view, God has full foreknowledge of the results of the evolutionary process that he sets in motion, making Him causally and morally responsible.[19]

And even if God did not have full foreknowledge—such as in an open theist account, in which the future is unknowable, since it does not yet exist—God would still have sufficient knowledge and power to be able to intervene to prevent the evolution of such undesirable features. It further also seems that God could do this without any damage to the possibility of biological organisms' self-determination of their own evolution, because if these organisms do not have libertarian free will, then they are not self-determining in any case. And if we adopt a different, compatibilist understanding of agency, with which self-determination is compatible with not being free in the libertarian sense, then this kind of free will would also be compatible with a divine determination of creatures' desires. Thus, while the idea of evolutionary theodicy based on creaturely autonomy and agency seems appealing, I remain unconvinced of the notions of agency and freedom that the argument depends on.

Perhaps it would also be possible to construct a type of evolutionary theodicy without reference to the value of freedom, simply by following the principle that God has left some room for real chance events within his processes of creation. As noted, many have argued for the intrinsic goodness of a process-like method of creation, as opposed to a miraculous creation. But then, the difficulty lies in establishing the plausibility that evolution should inevitably result in outcomes like parasitism, or that some value would be lost from the divine guidance of evolution to prevent such outcomes. The doctrine of creation out of nothing, as traditionally understood, would seem to imply that any evolutionary process is dependent on God. In this book, I have argued that the possibilities that are open to evolution are constrained and created by God, whether these possibilities are then freely chosen by creatures or attained by contingent evolutionary processes. Given this, can we truly affirm that it was inevitable that the evolution of biological teleology through a process of evolution would also result in undesirable phenomena like parasitism?

Based on the understanding of evolution outlined in this book, one possibility would be to argue that the prospect of evolving parasitism and predation is an inevitable by-product of the way the "library of forms" must be designed in order to allow for evolution at all, as long as chance is allowed a role in this process. In this understanding, the library of forms that God has designed allows for evolution to navigate through it by small mutations, eventually filling the majority of niches that it can get to. It may be that it is logically impossible to design a library of forms that would allow for evolution without also allowing for the evolution of parasitism and predation, while also giving some room for chance within the creation. We can see this again through considering the example of the evolution of technology. Malicious inventions typically build on technology that is also useful for other purposes. Thus, instruments of torture are typically based on inventions that are also used for beneficial purposes, merely combining existing technology in new ways to serve an evil intent. Similarly, perhaps the survival strategies of parasites, pathogens, and predation simply adapt divinely intended forms to unintended uses. Because of physics, there may not be any way to allow for the evolution of jaws, teeth, and limbs without also allowing for the evolution of sharp teeth that are useful for predation, and strong limbs that can be used to incapacitate other creatures. This evolutionary theodicy would also fit with the traditional theological view of evil as the lack of good, or as the corruption of originally good things.

As De Smedt and De Cruz (2020, 21) and McLeish (2020) note, many in the theistic traditions have historically allowed chance a role in creation, even before evolutionary theory. The idea that God determines evolutionary outcomes would only partly fit well with these ideas. The question remains why God does not intervene miraculously to prevent the evolution of at least the most harmful parasites and pathogens, if God did not intend for their existence directly. Even though the possibility of the evolution of such harmful ways of life might be a necessary consequence of the structure of any library of forms, the actualization of such forms is not thereby established as a necessity. Thus evolutionary theodicy still strikes me as incomplete—at least when coupled with a traditional understanding of divine attributes.[20] Nevertheless, the question "Why does God not intervene more, if he intervenes at all?" does similarly apply to all other views of divine providence that allow for divine interventions. And the response will typically be that we are not in a position to evaluate how much intervention actually occurs, and that God intervenes as much as is metaphysically, logically, and morally possible without compromising other goods—and will compensate the sufferers in the world to come (Schneider 2020).

I have argued that the formulation of the design argument (and design discourse) that I have defended here adds to the credibility of an evolutionary theodicy, rather than subtracting from it. The idea of evolutionary creation as revealing the Creator's glory provides a better account of why the Creator might choose to create through an evolutionary process, and the idea of the Creator as being responsible for the "library of forms" that underlies the possibility of evolution provides a way of showing how divine design could explain the existence of biological teleology, but simultaneously leave details up to chance—insofar as such a response is possible under any account of divine causal and moral responsibility in relation to creaturely freedom. It is possible to argue that this evolutionary theodicy fails, but then it would also fail for other forms of theistic evolutionism, and hence my defense of the biological design argument does not make the problem more difficult. However, even supposing that this theodicy fails, there is an extensive literature on the problem of natural evil, with many different responses. Indeed, it seems likely to me that a complex problem like the existence of natural evil cannot be explained by reference to just one factor (Kojonen 2016a, 162–167).

Therefore, I am content to conclude that insofar as an evolutionary theodicy is able to resolve at least part of the problem of natural evil and bad design, then such a theodicy can be used just as well by defenders of the biological design arguments, as long as these are understood in a way that is compatible with evolutionary theory. Evolutionary design arguments in the form that I have developed are additionally also compatible with other responses to the problems of natural evil and animal suffering, such as Neo-Cartesianism, Fall-based theodicies (referencing an angelic or human Fall, or both) and soulmaking theodicies. In emphasizing the dependency of evolution on the Creator, my argument does make it more difficult to assert that the course of evolution is entirely contingent. But it does allow room for enough contingency to make an evolutionary theodicy feasible. Furthermore, the non-randomness of evolution is an inference arising out of scientific data, and thus would need to be dealt with by proponents of evolutionary theodicies regardless of the salvaging of the design argument. The doctrines of creation out of nothing and divine foreknowledge would also lead to the conclusion that God must be responsible for any evolutionary process, and must thus have some responsibility also for the order of biological organisms.

Whereas some have argued that theology's main concern with biology should be to absolve God of the responsibility for biological bad designs, it seems to me that this should not be done in a way that also absolves God of the glory for his creation. It may be that it will ultimately prove possible to provide a balanced account of biological design that allows us to praise God's divine wisdom for the beauty and orderly complexity of biological organisms, while not blaming Him for apparently malicious features. Perhaps we already have such an argument here, in the salvaged biological design argument, combined with an evolutionary theodicy (with perhaps some cosmic conflict thrown into the mix). But in the absence of such an argument, it seems to me that it would be theologically preferable to give God both the glory and the blame, rather than to avoid giving Him both.[21]

5.4 Conclusions

Explanations are commonly understood as compatible if they do not answer the same question, but are instead answers to different, even if overlapping, questions. In that case, neither explanation will really be unnecessary. Moreover, explanations of even the same phenomenon can also be compatible if they form part of an explanatory chain, where the proximate cause is explained by another cause that stands behind this cause. I have argued that both of these strategies can be applied to the biological design argument, once the tight dependency of evolution on the wider teleology of the cosmos is acknowledged. Thus design would explain the general existence of biological teleology, while evolution would still be left to explain the particulars (at least if no further reasons are given for believing that God directs evolution beyond what is needed for the biological design argument). The revelatory potential of biological creation becomes particularly clear if we adopt an understanding of evolution where the course of evolution is highly constrained by "the laws of form." Even if we adopt a view that emphasizes the contingency of evolution, it still seems to be clear that the existence of evolutionary potential has tight preconditions. Thus the order of biological organisms can reveal the evolutionary potential and fecundity of the cosmos, and thus also cause us to wonder at this cosmos that we live in, and what kind of Creator it might have. Indeed, it would be odd if biological organisms as the product of a fine-tuned cosmos could not manifest the designed character of that cosmos reliably. Through many analogies and thought experiments, I have argued that creation through secondary causes can also allow for the end product to be revelatory of the Creator.

All this being said, I recognize that the account given here could be developed much further. For example, the credibility of the account depends on our understanding of the philosophy of explanation, and here much more work could be done to relate recent work in the study of causality and explanation to the philosophy of religion. There is also a potential link here to the theology of technology and human culture, which I have only briefly hinted at. We might want to say that facets of human technology and human culture might also reflect the Creator's greatness, even while we would not want to say that the Creator is responsible for the particular formulations of culture and the uses of technology. Concepts of causal and moral responsibility are central here.

I have argued that understanding biological teleology as providing the perception, sign, or evidence of design can also be justified within an evolutionary view of the cosmos. However, as with all perception, the worldview and vantage point of the observer will affect what is perceived. As I have argued, the design argument works best as part of a wider Christian theology of nature or a cumulative natural theology, in which propositions about what the Creator would or would not do can be supported by reference to other reasons. And this also means that we should confront the existence of evil, which forms the most common counterargument to theism generally and design arguments more specifically. The design argument can be understood as a statement of the "problem of good" that can at least counterbalance the problem of evil.

While many have argued that evolution makes the problem of evil worse, it seems to me that the combination of evolution and the biological design argument, at least, helps rather than hinders in responding to the problem of evil. First, there is the "problem of good" noted above as a counterbalance to the problem of evil, providing a more accurate assessment of the total evidence. Second, this version of the design argument can be combined with evolutionary theodicies, providing a framework which makes the existence of apparently malicious designs at least somewhat more intelligible. Third, this version of the design argument, not being a scientific argument, can be combined with a more robust Christian theology of nature, allowing for the interpretation of nature to be supplemented by a Christological and eschatological understanding. Finally, the arguments of natural theology can themselves help support belief that the Creator is greater and wiser than us, despite the suffering we experience.

If God has created through a process of designed evolution, as theistic evolutionism implies, then this also implies that creatures do not need to conform to our standards of engineering perfection before the order of biology can provide evidence of design. Insofar as evolution fills all niches that are available to it, it may be plausible to argue that God did not even directly intend all results of the process. Such an evolutionary theodicy leaves difficult open questions, particularly related to the causal and moral responsibility of the Creator in relation to the freedom of creatures. Nevertheless, this chapter has at least shown that, contrary to popular opinion, evolutionary theodicies are compatible with a design-based perspective holding that God is responsible for the existence of biological teleology.

And thus, contrary to some received views, affirming the overall designedness of biological nature does not need to make the task of theodicy more difficult. Indeed, in affirming that we have at least some sign of the Creator's wisdom, I suspect that a design-based perspective is more likely to assist us in responding to the problem. If we can recognize signs of the Creator's wisdom and benevolence in at least parts of nature, and if we can see some beauty in finely tuned laws of form that could lead to the evolution of biology's beauty from hummingbirds to human brains, then could this not provide at least a little support for faith that Providence might have some plan also to redeem the darker side of nature?

Notes

1. Kim's argument is applied to the case of theistic evolution by Leidenhag (2019), who argues that many forms of theistic evolution in which God's causal role is not clearly specified can indeed be criticized for causal over-determination. However, Leidenhag differs from ID proponents in arguing that the Creator's causal role does not need to be scientifically detectable. For an excellent discussion of these points, see Skogholt (2020) and Leidenhag (2020).
2. For a further, more robust, reconstruction of Mayr's concepts, see Ariew (2003). I am indebted to Visala (2011, 125–139).
3. Here, the role ascribed to God may seem too limited for some. For them, I would add that a wider theological perspective on biological purpose could also answer other questions, such as "Why does evolution appear to have a direction", and might provide further understanding of how divine causation relates to the particulars of evolution. For instance, one might have theological reasons for believing (or not believing) that divine control of evolution extends beyond that required for setting up the process.
4. It seems to me that this duality of possible design arguments may create a dilemma for those who would rely on evolution as a defeater of design arguments. The proponent of design could argue roughly as follows: if evolution is highly contingent and its outcome is highly improbable, then reaching this outcome would be better explained by theism. However, if evolution is not as improbable and not as contingent, then the patterns of evolution are explained to a large degree by the background conditions and laws, and evolution simply pushes the search for an explanation further.
5. To be fair, there is a way to extend Plantinga's account in a manner close to the idea that the divine guidance of mutations could help supplement evolutionary explanations, mentioned at the end of this chapter. Plantinga could argue that we can accept evolutionary explanations as correct, but that taken by themselves evolutionary explanations are still improbable (or cannot predict the actual evolution of a species like humans), and hence this leaves space for supplementing evolutionary explanations with the additional idea of divinely guided mutations. Nevertheless, evolutionary explanations are not typically regarded as this improbable.
6. The possibilities of commonsense reasoning as the basis of design arguments is also defended by Axe (2017). However, unlike the strategy adopted here, Axe's defense is anti-evolutionist.
7. Vodder (2018) pressed me for clarity on this point in response to Kojonen (2016b), in which I presented an earlier version of some arguments here.
8. The program discussed by Wahlberg is Tonica Fugata 9.0., released by Capella Software, but many other similar programs also exist. A free demo

version can be downloaded from http://www.capella.de/Download_tonica.cfm.

9. Harrison (2017) is an excellent discussion of the metaphysics of mathematics and its relevance for theism.
10. Ayala's theodicy is critically analyzed by Doran (2009) and Berhow (2019).
11. In the philosophy of religion, such a "problem of good" has recently been used in the discussion of the "evil God" hypothesis (Page and Baker-Hytch 2020).
12. Quoted in Loikkanen (2015, 181).
13. This, at least, is Dembski's position as he has clarified it in dialogue with his critics, such as Michael Ruse. Dembski (2004, 9) argues that "Ruse is wrong that the Explanatory Filter separates necessity, chance and design into mutually exclusive and exhaustive categories." See also Dembski (2008, 221) for similar comments. As Corabi (2009) and Berhow (2019) also note, the ID perspective allows for a great deal of regularity in nature.
14. Baker-Hytch (2019) argues that harms may be connected in a way that if are able to suffer, then it is better for them to be mortal. Peels' (2018) analysis of natural selection fits well with this kind of argument.
15. Wahlberg aims his critique particularly against the versions of the evolutionary theodicy presented by Haught (2000, 53), Polkinghorne (2011, 83), and Southgate (2008).
16. Quoted in Murray 2011, 144–145.
17. There have been attempts to argue that the human fall might have effects on the history of the cosmos already before the Fall. See, for example, Dembski (2008); for a critique see Sollereder (2018, 67–70) and Peels (2018, 552–554).
18. Murray (2011, 170–175) presents further problems for the idea that evolution is necessary for moral autonomy, or that creation out of nothing would threaten it.
19. Here my discussion, however, sidesteps the major theological views of compatibilism and theological determinism—some forms of which attempt to argue that divine determinism does not remove creaturely culpability. The link between responsibility and freedom was notably criticized by the Protestant reformers Luther (Visala, Vainio 2020), and Calvin (e.g., Bignon 2017). Many of the same theologians would also oppose the application of the concept of moral responsibility to God, and our ability to evaluate God's actions in general (Davies 2006; Dougherty and McBrayer eds. 2014).
20. Adopting a full-blown process view of God, in which divine omnipotence and creation out of nothing are questioned, might arguably provide a different picture, but would also mean giving up central features of the traditional understanding of God, which would result in other serious problems.

For example, as Pak (2014, 2016) argues in depth, in such a view it is difficult to guarantee that good will ultimately triumph over evil.

21. While it may be difficult to find a line of argument that gives God only the glory but not the blame, I think there is a stronger difficulty in giving God only the blame but not the glory. If God's use of evolution does not remove God's responsibility for the bad features that evolve (despite creaturely freedom), then why should it remove God's responsibility for the good features that evolve? It therefore seems to me that the rejection of the evolutionary theodicy as unworkable should also lead to the rejection of evolutionary explanations as removing divine glory for biology.

References

Argyll, George. 1885, April. What Is Science? *Good Words*, 236–245.

Ariew, André. 2003. Ernst Mayr's 'Ultimate/Proximate' Distinction Reconsidered and Reconstructed. *Biology and Philosophy* 18: 553–565.

Axe, Douglas. 2017. *Undeniable: How Biology Confirms Our Intuition That Life Is Designed*. San Francisco, CA: HarperOne.

Ayala, Francisco. 2007. *Darwin's Gift to Science and Religion*. Washington, DC: Joseph Henry Press.

Baker-Hytch, Max. 2019. God, Evil and Organic Wholes. Unpublished Paper, Presentation at the Helsinki Analytic Theology Workshop 2019.

Behe, Michael J. 1999. Review. God After Darwin: A Theology of Evolution. *Metaviews*, December 4.

———. 2007. *The Edge of Evolution: The Search for the Limits of Darwinism*. New York, NY: The Free Press.

Berhow, Michael. 2019. *Dysteleology: A Philosophical Assessment of Suboptimal Design in Biology*. Eugene, OR: Wipf and Stock.

Bhaskar, Roy. 1978. *A Realist Theory of Science*. 2nd ed. Hassocks: Harvester Press.

Bignon, Guillaime. 2017. *Excusing Sinners and Blaming God: A Calvinist Assessment of Determinism, Moral Responsibility, and Divine Involvement in Evil*. Eugene, OR: Pickwick.

Brooke, John Hedley. 1991. *Science and Religion: Some Historical Perspectives*. Cambridge: Cambridge University Press.

Collins, Robin. 2018. The Argument from Physical Constants: Fine-Tuning for Discoverability. In *Two Dozen (Or So) Arguments for God's Existence*, ed. Jerry L. Walls and Trent Dougherty, 89–107. Oxford: Oxford University Press. A Previous Draft Is Available at https://pdfs.semanticscholar.org/6ab2/7366b32edcf3d12587a07397e0d1d8e496fd.pdf.

Conway Morris, Simon, and Stephen Jay Gould. 1998. Showdown on the Burgess Shale. *Natural History Magazine* 107 (10): 48–55. Available at http://www.stephenjaygould.org/library/naturalhistory_cambrian.html.

Corabi, Joseph. 2009. Intelligent Design and Theodicy. *Religious Studies* 45 (1): 21–35.

Cunningham, Conor. 2010. *Darwin's Pious Idea: Why the Ultra-Darwinists and Creationists Both Get It Wrong*. Grand Rapids, MI: William B. Eerdmans.

Darwin, Charles. 1860. To Asa Gray, 22 May 1860. *Darwin Correspondence Project*, Letter No. 2814. Available at https://www.darwinproject.ac.uk/letter/DCP-LETT-2814.xml.

———. 1958 [1887]. In *The Autobiography of Charles Darwin*, ed. Nora Barlow. London: Collins.

Davies, Brian. 2006. *The Reality of God and the Problem of Evil*. London and New York, NY: Continuum.

Dawkins, Richard. 2006. *The God Delusion*. London: Bantam Press.

De Smedt, Johan, and Helen De Cruz. 2020. *The Challenge of Evolution to Religion*. Cambridge: Cambridge University Press.

Dembski, William A. 1999. *Intelligent Design: The Bridge Between Science and Theology*. Downers Grove, IL: InterVarsity Press.

Dembski, William. 2002. *No Free Lunch: Why Specified Complexity Cannot Be Purchased Without Intelligence*. Lanham: Rowman & Littlefield Publishers.

Dembski, William A. 2004. *Design Revolution: Answering the Toughest Questions About Intelligent Design*. Downer's Grove, IL: Intervarsity Press.

———. 2008. Making the Task of Theodicy Impossible? Intelligent Design and the Problem of Evil. In *The Evolution of Evil*, ed. Gaymon Bennett, Martinez J. Hewlett, Ted Peters, and Robert J. Russell, 218–233. Göttingen: Vanderhoeck & Ruprecht.

Dennett, Daniel C. 1995a. *Darwin's Dangerous Idea: Evolution and the Meanings of Life*. New York, NY: Simon & Schuster.

Dennett, Daniel. 1995b. Comment on Brian Goodwin, 'Biology is Just a Dance.' *Edge Conversations*. Edge.org. Available at https://www.edge.org/conversation/brian_goodwin-chapter-4-biology-is-just-a-dance#21968.

Dennett, Daniel, and Alvin Plantinga. 2011. *Science and Religion: Are They Compatible?* Oxford: Oxford University Press.

Dickins, T., and R. Barton. 2013. Reciprocal Causation and the Proximate–Ultimate Distinction. *Biology & Philosophy* 28 (5): 747–756.

Doran, Chris. 2009. From Atheism to Theodicy to Intelligent Design: Responding to the Work of Francisco J. Ayala. *Theology and Science* 7 (4): 337–344.

Dougherty, Trent. 2014. *The Problem of Animal Pain: A Theodicy for All Creatures Great and Small*. Basingstoke: Palgrave.

Dougherty, Trent, and Justin P. McBrayer, eds. 2014. *Skeptical Theism: New Essays*. Oxford: Oxford University Press.

Draper, Paul. 2012. Darwin's Argument from Evil. In *Scientific Approaches to the Philosophy of Religion*, ed. Yujin Nagasawa, 49–70. Basingstoke: Palgrave Macmillan.

Fergusson, David. 2018. *The Providence of God: A Polyphonic Approach*. Cambridge: Cambridge University Press.

Feser, Edward. 2017. *Five Proofs of the Existence of God*. San Francisco, CA: Ignatius Press.

Garvey, Jon. 2019. *God's Good Earth: The Case for an Unfallen Creation*. Eugene, OR: Cascade Books.

George, Marie. 2002. On Attempts to Salvage Paley's Argument from Design. In *Science, Philosophy, Theology*, ed. John O'Callaghan. South Bend, IN: St. Augustine's Press.

Glass, David. 2012. Can Evidence for Design be Explained Away? In *Probability in the Philosophy of Religion*, ed. V. Harrison and J. Chandler, 79–102. Oxford: Oxford University Press.

Glass, David, and Mark McCartney. 2014. Explaining and Explaining Away in Science and Religion. *Theology and Science* 12 (4): 338–361.

Glass, David, and Schupbach, Jonah. 2020. Conjunctive Explanations. Unpublished Paper, Part of the Conjunctive Explanations in Science and Religion Project (2018–2020).

Gould, Stephen Jay, and Richard Lewontin. 1979. The Spandrels of San Marco and the Panglossian Paradigm: A Critique of the Adaptationist Programme. *Proceedings of the Royal Society. Series B* 205 (1161): 581–598.

Gray, Asa. 1860. Natural Selection Not Inconsistent with Natural Theology. *Atlantic Monthly for July, August and October*, 1860. *Darwiniana: Essays and Reviews Pertaining to Darwinism*. New York: D. Appleton, 1888 [1874].

Harrison, Victoria S. 2017. Mathematical Objects and the Object of Theology. *Religious Studies: An International Journal for the Philosophy of Religion* 53: 479–496.

Hart, David Bentley. 2005. *The Doors of the Sea: Where Was God in the Tsunami?* Grand Rapids, MI: Wm. B. Eerdmans Publishing Co.

Haught, John F. 2000. *God After Darwin: A Theology of Evolution*. Boulder, CO: Westview Press.

Henrich, Joseph. 2015. *How Culture Is Driving Human Evolution, Domesticating Our Species, and Making Us Smarter*. Princeton, NJ: Princeton University Press.

Johnson, Phillip E. 1993. Creator Or Blind Watchmaker? *First Things* 29: 8–14.

Johnson, Curtis. 2015. *Darwin's Dice: The Idea of Chance in the Thought of Charles Darwin*. New York, NY and Oxford: Oxford University Press.

Kim, Jaegwon. 2007. *Physicalism, Or Something Near Enough*. Princeton, NJ: Princeton University Press.

Kojonen, Erkki Vesa Rope. 2016a. *The Intelligent Design Debate and the Temptation of Scientism*. London: Routledge.

———. 2016b. Salvaging the Biological Design Argument in Light of Darwinism? *Theology and Science* 14 (3): 361–381.

———. 2018. Design Discourse: A Way Forward for Theistic Evolutionism. *Neue Zeitschrift für Systematische Theologie und Religionsphilosophie* 60 (3): 435–451.

———. 2019. Is Classical Science in Conflict with Belief in Miracles? Some Bridge-Building between Different Positions. In *God's Providence and Randomness in Nature: Scientific and Theological Perspectives*, ed. Robert John Russell and Joshua M. Moritz, 205–234. West Conshohocken, PA: Templeton Press.

Laland, Kevin, Kim Sterelny, John Odling-Smee, William Hoppitt, and Tobias Uller. 2011. Cause and Effect in Biology Revisited: Is Mayr's Proximate–Ultimate Dichotomy Still Useful? *Science* 334 (6062): 1512–1516.

Lamont, John. 2004. *Divine Faith*. Aldershot: Ashgate.

Leibniz, Gottfried. 1956. In *The Leibniz-Clark Correspondence*, ed. H.G. Alexander. Manchester: Manchester University Press.

Leidenhag, Mikael. 2019. The Blurred Line Between Theistic Evolutionism and Intelligent Design. *Zygon* 54 (4): 909–931.

———. 2020. The Problem of Natural Divine Causation and the Benefits of Partial Causation: A Response to Skogholt. *Zygon* 55 (3): 696–709.

List, Christian. 2019. *Why Free Will Is Real*. Cambridge, MA: Harvard University Press.

Loikkanen, Juuso. 2015. Early Lutheranism and Natural Theology. *Teorie Vědy/ Theory of Science* 37 (2): 173–186.

Luther, Martin. 1921 [1529]. The Large Catechism. In *Triglot Concordia: The Symbolical Books of the Evangelical Lutheran Church*. German-Latin-English. St. Louis: Concordia.

———. 1964. Lectures on Genesis. In *Luther's Works*, ed. Jaroslav Pelikan, vol. 6. St. Louis: Concordia.

Mayr, Ernst. 1961. Cause and Effect in Biology. *Science* 134 (3489): 1501–1506.

McGrath, Alister. 2002. *A Scientific Theology, Vol. 2: Reality*. London: T&T Clark.

———. 2009. *Darwinism and the Divine: Evolutionary Thought and Natural Theology*. Oxford: Wiley-Blackwell.

———. 2016. *Re-Imagining Nature: The Promise of a Christian Natural Theology*. Oxford: Wiley-Blackwell.

McLeish, Tom. 2014. *Faith and Wisdom in Science*. Oxford: Oxford University Press.

———. 2020. Evolution as the Unwrapping of the Gift of Freedom. *Scientia et Fides* 8 (2): 43–64. https://doi.org/10.12775/SetF.2020.014.

McMullin, Ernan. 2009. On the Origin of Terrestrial Life: A Christian Perspective. In *Exploring the Origin, Extent, and Future of Life: Philosophical, Ethical, and Theological Perspectives*, ed. Constance M. Bertka, 80–95. Cambridge: Cambridge University Press.

Meyer, Stephen C. 2017. The Difference It Doesn't Make: Why the 'Front-End Loaded' Concept of Design Fails to Explain the Origin of Biological Information. In *Theistic Evolution: A Scientific, Philosophical and Theological*

Critique, ed. J.P. Moreland, Stephen C. Meyer, Christopher Shaw, Ann K. Gauger, and Wayne Grudem, 217–236. Wheaton, IL: Crossway.

Miller, Calum. 2020. Do Animals Feel Pain in a Morally Relevant Sense? Philosophia. Published Online 13 October. https://doi.org/10.1007/s11406-020-00254-x.

Moritz, Joshua M. 2019. Contingency, Convergence, Constraints, and the Challenge from Theodicy in Creation's Evolution. In *God's Providence and Randomness in Nature: Scientific and Theological Perspectives*, ed. Robert J. Russell and Joshua M. Moritz, 289–328. West Conshohocken, PA: Templeton Press.

Murray, Michael J. 2011. *Nature Red in Tooth and Claw: Theism and the Problem of Animal Suffering*. Oxford: Oxford University Press.

Nagel, Thomas. 2012. *Mind and Cosmos Why the Materialist Neo-Darwinian Conception of Nature Is Almost Certainly False*. New York, NY and Oxford: Oxford University Press.

Nowak, Michael, and Sarah Coakley, eds. 2013. *Evolution, Games, and God: The Principle of Cooperation*. Cambridge, MA: Harvard University Press.

Oord, Thomas Jay. 2015. *The Uncontrolling Love of God: An Open and Relational Theology*. Downers Grove, IL: IVP Academic.

Page, Ben, and Max Baker-Hytch. 2020. Meeting the Evil God Challenge. *Pacific Philosophical Quarterly* 101 (3): 489–514.

Pak, Kenneth. 2014. Could Process Theodicy Uphold the Generic Idea of God? *American Journal of Theology and Philosophy* 35 (3): 211–228.

———. 2016. *Divine Power and Evil: A Reply to Process Theodicy*. London: Routledge.

Paley, William. 2008 [1802]. *Natural Theology*. Oxford World's Classics. Oxford: Oxford University.

Peckham, John. 2018. *A Theodicy of Love: Cosmic Conflict and the Problem of Evil*. Grand Rapids, MI: Baker Publishing.

Peels, Rik. 2018. Does Evolution Conflict with God's Character? *Modern Theology* 34 (4): 544–564.

Persson, Johannes, and Petri Ylikoski, eds. 2007. *Rethinking Explanation*. Dordrecht: Springer.

Peters, Ted, and Martinez Hewlett. 2003. *Evolution from Creation to New Creation: Conflict, Conversation and Convergence*. Nashville, TN: Abingdon Press.

Plantinga, Alvin. 1977. *God, Freedom, and Evil*. Grand Rapids, MI: Wm. B. Eerdmans Publishing Company.

———. 2011. *Where the Conflict Really Lies: Science, Religion, and Naturalism*. New York, NY: Oxford University Press.

Polkinghorne, John. 2011. *Science and Religion in Quest of Truth*. London: Society for Promoting Christian Knowledge.

Ratzsch, Del. 2001. *Nature, Design and Science: The Status of Design in Natural Science*. Albany, NY: State University of New York Press.

Ratzsch, Del, and Koperski, Jeffrey. 2019. Teleological Arguments for God's Existence. *The Stanford Encyclopedia of Philosophy* (Summer 2019 Edition). Ed. Edward N. Zalta. Available at https://plato.stanford.edu/archives/sum2019/entries/teleological-arguments/.

Russell, Robert J. 2013. Recent Theological Interpretations of Evolution. *Theology and Science* 11 (3): 169–184.

Schaffer, Jonathan. 2016. The Metaphysics of Causation. *The Stanford Encyclopedia of Philosophy* (Fall 2016 ed.). Ed. Edward N. Zalta. Available at https://plato.stanford.edu/archives/fall2016/entries/causation-metaphysics/.

Schmidt, William E. 1991. 2 'Jovial Con Men' Demystify Those Crop Circles in Britain. *The New York Times*, September 10.

Schneider, John R. 2020. *Animal Suffering and the Darwinian Problem of Evil*. Cambridge: Cambridge University Press.

Schupbach, Jonah N. 2016. Competing Explanations and Explaining-Away Arguments. *Theology and Science* 14 (3): 256–267.

Sedley, David. 2007. *Creationism and Its Critics in Antiquity*. Sather Classical Lectures 66. Berkeley, CA: University of California Press.

Skogholt, Christoffer. 2020. I Walk the Line: Comment on Mikael Leidenhag on Theistic Evolution and Intelligent Design. *Zygon* 55 (3): 685–695.

Snellman, Lauri. 2019. Anti-theodicy and Antitheodicies. *European Journal for Philosophy of Religion* 11 (1): 201–211.

Sollereder, Bethany. 2018. *God, Evolution, and Animal Suffering: Theodicy Without a Fall*. London: Routledge.

Southgate, Christopher. 2008. *The Groaning of Creation: God, Evolution, and the Problem of Evil*. Louisville, KY: Westminster John Knox Press.

———. 2018. *Theology in a Suffering World: Glory and Longing*. Cambridge: Cambridge University Press.

Sterba, James P. 2019. Solving Darwin's Problem of Natural Evil. *Sophia* 59: 501–512.

Stump, Eleonore. 2010. *Wandering in Darkness: Narrative and the Problem of Suffering*. Oxford: Clarendon.

Talbert, Matthew. 2016. *Moral Responsibility: An Introduction*. Cambridge: Polity Press.

Timpe, Kevin. 2013. *Free Will in Philosophical Theology*. New York, NY and London: Bloomsbury.

Tomasello, Michael. 2019. *Becoming Human: A Theory of Ontogeny*. Cambridge, MA: Harvard University Press.

Tracy, Thomas F. 2013. Divine Purpose and Evolutionary Processes. *Zygon* 48 (2): 454–465.

Visala, Aku. 2011. *Theism and the Cognitive Science of Religion*. Farnham: Ashgate.

———. 2019. *Vapaan Tahdon Filosofia*. Helsinki: Gaudeamus.

Visala, Aku, and Olli-Pekka Vainio. 2020. Erasmus versus Luther: A Contemporary Analysis of the Debate on Free Will. *Neue Zeitschrift für systematische Theologie und Religionsphilosophie* 62 (3): 311–335.

Vodder, Hans W. 2018. Salvaging Biological Design Or Doing Away with It? A Criticism of Kojonen's Proposal. Unpublished Paper.

Wagner, Andreas. 2016. Without a Platonic Library of Forms, Evolution Couldn't Work. *Aeon.com*. March 16. Available at https://aeon.co/essays/without-a-library-of-platonic-forms-evolution-couldn-t-work.

Wahlberg, Mats. 2012. *Reshaping Natural Theology: Seeing Nature as Creation*. London: Palgrave Macmillan.

———. 2015. Was Evolution the Only Possible Way for God to Make Autonomous Creatures? Examination of an Argument in Evolutionary Theodicy. *International Journal for Philosophy of Religion* 77 (1): 37–51.

———. 2017. A Cosmological Argument Against Physicalism. *European Journal for Philosophy of Religion* 9 (2): 165–188.

Wigner, Eugene. 1960. The Unreasonable Effectiveness of Mathematics in the Natural Sciences. Richard Courant lecture in Mathematical Sciences Delivered at New York University, May 11, 1959. *Communications on Pure and Applied Mathematics* 13: 1–14.

Wiker, Benjamin, and Jonathan Witt. 2006. *A Meaningful World: How the Arts and Sciences Reveal the Genius of Nature*. Downers Grove, IL: IVP Academic.

CHAPTER 6

Conclusion: From Teleology to Theology, and Back Again

In this book, I have critically examined the traditional idea of conflict between design arguments and evolution. Against this idea, I have argued that the apparent teleology of biology continues to testify to or provide evidence of design, even in an evolutionary cosmos. It seems to me that taking this idea seriously has the potential to transform present discussion of the relationship between creation and evolution.

Historically, the idea of the compatibility of evolution and design arguments is not unprecedented, though it has been largely neglected in our contemporary discussion. Particularly during the Darwinian revolution, many authors presented the view that the evolutionary view provides us with better evidence of design than the view of the miraculous creation of species. I have already quoted Charles Kingsley to this effect; Henry Ward Beecher similarly argues that a rug does not manifest any less design if it is created by a rug-making machine rather than being sewed directly by a seamstress. According to Beecher (1885, 116), "design by wholesale is grander than design by retail."[1] It is somewhat unclear why this perspective of biological design has fallen by the wayside historically, and why so many have held to the view that evolution and design should be seen as competing explanations. Ruse (2018) has documented the widespread use (or abuse) of evolutionary theory as the foundation of a secular religion; perhaps the reasons for disregarding the perspective of design are tied to this historically contingent development. It may also be that the

E. V. R. Kojonen, *The Compatibility of Evolution and Design*, Palgrave Frontiers in Philosophy of Religion, https://doi.org/10.1007/978-3-030-69683-2_6

widespread influence of the critique of traditional natural theology and the influence of scientistic and creationist thinking have also contributed to overlooking this type of design argument. The influence of Darwin's critique of Asa Gray and the desire for more distance between theology and the natural sciences are also factors (Johnson 2015). There were sociological and cultural forces in play against such a harmonization (Stanley 2015). , and the demise of design arguments is historically correlated with the rise of a positivistic philosophy of science (Cashdollar 1989). In the intellectual environment of the late nineteenth and early twentieth centuries, which increasingly saw philosophical, theological and metaphysical claims as nonsensical, it is not surprising that accounts of the compatibility of design and evolution should fall out of favor. I think such ultimately nonrational currents of intellectual fashion do partly explain why evolutionary design arguments were forgotten. But it should also be noted that the scientific evidence for fine-tuning and directionality in evolution, which seem to vindicate Beecher's intuition, is also now better developed.

I have argued for the compatibility of design and evolution both by developing a philosophical account of how these explanations fit together and through thought experiments showing that indirect design, even through a process incorporating chance-based processes, need not result in lack of evidence of design. Though I think the philosophical formulation of the compatibility of evolution and design as complementary explanations succeeds, I am still more convinced of the overall intuitive conclusion of compatibility than of the exact philosophical version of the argument that I have given. Examples like Beecher's rug factory, Ratzsch's automated VCR factory, genetic algorithms, music-generating algorithms, and architecture all seem to point to the conclusion that a designer's existence and wisdom can be manifested in the products of an indirect, even automated, process, and even in the products of a process that allows a substantial causal role for chance. I recognize that much further work can be done in the philosophical formulation of the argument, and overall in relating the newest philosophical analyses of explanations to theism. Once further work on understanding the nature of conjunctive explanations is done, the argument for the compatibility of evolution and design can no doubt be formulated in a more robust manner, as well.

I have argued that the activation of our intuitive, commonsense design detection in the case of biology does provide even the "theist on the street," (Shaw 2019) who does not know about the specifics of evolutionary biology, with a strong reason to suspect that an undesigned evolutionary process could not produce the marvels of teleology that we do observe. If this is correct, then those popularizers of evolutionary biology who have emphasized the unguided, unintelligent nature of evolution may be partly responsible for the continued public mistrust of evolutionary biology. It would typically be unreasonable to trust oneself over an expert in another field. However, there are limits to our ordinarily valid trust in epistemic authority. As Elga (2007, 483) notes, "upon finding out that my forecaster is confident that it will snow tomorrow, I will follow suit. But upon finding out that my forecaster is confident that it will rain eggplants tomorrow, I will not follow suit. I will conclude that my forecaster is crazy."[2] It seems to me that by using their faculties of design detection, ordinary theists on the street are able to recognize that Dawkins and others are stepping beyond mere biology in asserting that evolution works without design. Since Dawkins holds unguidedness to be a central tenet of evolutionary biology, this then ends up casting suspicion on the entire theory, though in fact in a more accurate understanding, evolution is compatible with design. Of course, it may be that for others, such metaphysical interpretations of evolution make the theory more attractive as the foundation of the kind of secular religion that Ruse describes. But the salvaging of the biological design argument opens up the possibility of theistic evolution for those of us who experience a strong intuition of designedness when observing living beings.

I have defended the *prima facie* reliability of our commonsense design detection. My hope is that in future treatments, this can be explored further by incorporating more analysis of the psychology and philosophy of perception (Wahlberg 2012; Clark and Barrett 2011; McGrath 2008). Evolution, understood as a refutation of design, has been the most important critique of the reliability of this human faculty when applied outside the domain of human artifacts. However, there continue to be other ways to criticize the reliability of the conclusion of design. For instance, some argue that detecting design always requires strong prior knowledge of the designer. And my analysis also supports the contention that belief in God (and belief that God does not exist) affects the credibility of the conclusion of design. However, I have also argued that it is epistemically undesirable to formulate our principles of rationality in such a way that evidence of design is already ruled out on a priori philosophical grounds. This would effectively commit us to a naturalism of the gaps, closing off the possibility of any evidential support for belief in God already before the investigation of the evidence. It seems that such a naturalism of the gaps should be avoided just as fervently as Gods of the gaps are avoided.

This formulation of the argument might also be criticized for moving the discussion over design partly into the realm of philosophy, metaphysics, and theology. Though the design argument is impacted by scientific results, I have argued that since it relies on personal explanation, it is fundamentally different from many explanations used in the natural sciences. This may be problematic for some. Hull (1973), for example, thinks that the pious harmonizations of teleology and evolution by nineteenth-century thinkers like Beecher and Kingsley in fact contributed to the demise of teleology, as does Pearcey (1998, 84). According to Hull (1973, 65), "What these men did not realize was that by pushing God further and further into the background as the unknowable author of natural law, [...] they had prepared the way for his total expulsion."[3] Nevertheless, it seems to me that this type of rejection of metaphysics depends on a scientistic and positivistic view of knowledge, and the consequent inattentiveness to the question of what ultimately explains the orderliness of nature. It assumes that belief in indirect design could not be rational and justified. But establishing this would require criticizing the premises of the argument, rather than simply assuming that the appearance of design could only provide support for religious belief if it was produced by supernatural miracles.

However, Hull's argument does show a way for critics of design arguments to also welcome the suggestion that evolution by itself does not refute the biological design argument, but moves discussion of it into philosophy. This is because, for them, to move a question in the nebulous realm of philosophy and metaphysics is to remove the question from rational consideration, and thus this move would neutralize design arguments as a threat against naturalism. But other critics of design arguments might be interested in making this distinction simply as a matter of accurately understanding the implications of evolution. Loesberg (2007, 122–123), for example, hints in this direction by arguing that proponents of ID and their critics often make the same mistake:

> Natural selection, as Darwin proposed it, explained how organisms have evolved into different species and the particular way that they have evolved. To do so, he certainly had to provide natural process that would produce complex biological features. But he did not have to assert, nor, Kant argued, could he have proven, that that natural process either was or was not an exhaustive explanation for complexity. The ambition to provide such exhaustive explanations pushes Intelligent Design but also Darwinian controversialists who think they can show Intelligent Design to be scientifically incorrect as opposed to simply not relevant to scientific discourse, beyond science into metaphysics.

I would say that it goes too far to state that design has no relevance for science, however. For instance, the account of design given here could in principle motivate further scientific research into questions like the amount of contingence in the evolutionary process and the fine-tuning of the genotype-phenotype map. This is also one of the points on which my argument here could be criticized. If we think that the structuralist understanding of evolution is correct, then it is easy to see how the evolution of biological forms is, to a large extent, moved back to the laws of physics, the explanation of which lies outside the sciences, and into the realm of metaphysics and theology. I have indeed argued that even without the strong sense of directionality that this kind of view seems to entail, the tight dependency of evolution on the wider teleology of the cosmos is still sufficient to cast doubt on evolution as a refutation of the design argument. Even in that case, evolved biological creatures could still manifest the evolutionary potential of the cosmos, and by extension the wisdom of the Creator. Nevertheless, in principle it should also be possible to argue against the necessity of fine-tuning for evolution. The precise extent of biological fine-tuning, and the fine-tuning of the library of forms, is still debatable. John Polkinghorne pointed out in 2000 that

> Our understanding of the nature and behaviour of complex systems is rudimentary; our insights into the pathways by which life developed on Earth are contentious and unsure; the problems involved in the full understanding of the functioning of even a single cell are immense. (Polkinghorne 2000, 74)

Though our understanding of evolution has developed greatly in the twenty years since Polkinghorne wrote this, the basic point about the incompleteness of our understanding remains true today. Today it seems that the case for directionality in evolution, and the inbuilt potentiality of creation, is stronger than in Polkinghorne's day, so the trajectory of the development of science seems to be in favor of the argument presented here. At least we might say with Alexander (2018, 167) that "far from looking stochastic and random, evolution looks highly organized and constrained—predictable to some extent, perhaps even with inevitable outcomes. […] these kinds of data—and many more—render the claim that evolutionary history is necessarily a purposeless history less plausible." And we can also say that the arguments here render the idea that evolution refutes the design argument, and that it refutes the rationality of an intuitive design discourse, far less plausible as well. My hope is that future scientific research will give us an even clearer picture of factors like the structure of a genotype-phenotype map, the precise ratio of functional and nonfunctional forms in theoretical morphospace, and the extent to which features like this affect the course of evolution.

Today practically all biologists would affirm the thesis that the direction that evolution takes is not wholly contingent, but is constrained by physics and chemistry. This means that not every world containing natural selection and hereditary variations would also include the kind of evolution we see. But biologists differ on how surprising these features of our world are, and how they affect evolution. Some even argue for the importance of a platonic "library of forms" underlying the possibility of evolution. Nevertheless, the more ambitious thesis of the directionality of evolution is still not universally accepted, and future discoveries could in principle still radically alter our understanding of the relative importance of contingency and constraint for evolution. Writing as a theologian, I have therefore attempted to provide arguments that could be used given a wide variety of ways to understand evolution. It seems to me that the problems of contingency and design as metaphysical problems are difficult to evade entirely: If evolution is directional, as posited by some, then this directionality is contingent on the laws and constants of nature allowing this directionality. If evolution is highly contingent, so that running the "tape of life" again would likely cause a very different result, then this makes it surprising that such valuable outcomes have in practice been reached. It is an exciting time to be following developments in evolutionary biology. If current trends continue, it seems that the case for the compatibility of design and evolution will become stronger as our understanding increases. Moreover, here I have focused on the traditionally central factors of natural selection and random mutation, since these have historically been the most important for responding to the design argument, and are widely considered as the most hostile to teleology. If the idea of biological organisms as revelatory of divine design is compatible with these mechanisms, then it should be even more easily compatible with the additional mechanisms included in the extended evolutionary synthesis.

One feature emphasized in the extended synthesis is the ability of creatures to influence their own evolution, such as in niche construction. Thus, the discovery of directionality and design in evolution also raises questions on how we should relate the freedom and autonomy of creatures to the directionality of the process. The discussion over how predictable and directional evolution is reveals differences in intuitions about the spiritual significance of directionality. As McGhee (2011, 271) writes, spirituality (and support for both theism and atheism) has been found both in the view that "evolution was unpredictable, contingent, nonrepeating, and directionless" and in the view that "evolution is predictable, preordained, and inevitable":

> It is clear that many who believe that evolution is predictable, preordained, and inevitable find comfort in the lawlike certainty of this point of view. Others react with horror to this point of view, and find it depressingly deterministic, stultifying rather than creative, enslaving rather than liberating. Where is individual freedom, where is creativity, in an evolutionary process that already has a predetermined destination?

The salvaged design argument outlined in this book aims to provide an option that allows for both contingency and directionality. It is sufficient for the argument that design explains the general possibility and existence of biological teleology, with the details left up to contingent and free processes. However, it remains unclear how free the processes of nature really are, and how they relate to the actions of God. This will depend on our general view of the relationship of God and nature, and what we think about moral and causal responsibility, as well as the compatibility of real freedom and determinism. As McLeish (2014, 183) notes, the question of balancing chaos and order goes back to the biblical material. Some favor a theological determinism in which God determines all, while others favor a view in which God allows for creatures to act with him as co-creators of the cosmos. It seems to me that the salvaged design argument is compatible with either of these. To increase the degree of determinism in the view, one could argue that God determines creatures' acts, and invisibly directs the mutations involved in the evolutionary process. To increase the degree of contingency, we would need to defend the freedom of all creatures to act autonomously and thus affect the direction of evolution. Both of these options fit with the argument of this book. I understand the natural sciences as presently offering only partial descriptions of reality, and thus they do not rule out this further divine action (Kojonen 2019).

In the future, there is still much room to expand this account theologically, and what has been said here could be related to the Christian tradition in much more depth. For instance, regarding determinism, while analogies of factories and machines easily tend toward the deterministic view, the history of theology allows for a richer set of metaphors of the God—world relationship. These could help in more clearly distinguishing the directionality required by the design argument from determinism. I have already noted the examples of the cosmos as a temple or a house; another possibility would be to conceive of the cosmos using the biblical metaphors of a field or a garden, where God as the farmer or gardener has set up conditions which allow for the cultivation of teleology, but also allow for quite a bit of wildness. The metaphor of the garden also calls to mind the plausible existence of teleology outside of biology as well, even extending to basic physics (Simpson 2021).

I have argued that the salvaging of the design argument provides theological value in preserving the validity of intuitive design detection in nature, and retaining the idea of divine design as part of the doctrine of creation. Divine design is not equivalent to human design, but neither should it be understood as wholly disanalogous to it. I have also argued that this approach aids in recognizing signs of divine wisdom that are all around us, and it even helps in responding to the problem of evil. Here too the tradition has interesting precedents for developing the argument further in the specific context of a Christian theology of nature. For example, as Southgate (2008, 61) argues, for "Maximus the Confessor, everything has meaning through its participation in the Logos." Maximus' theology of created beings as reflecting the mind of God allows, according to Southgate, for just the kind of fluid understanding of creaturely essences that evolution requires (Blowers 2016). Southgate (2008, 61n40) argues that:

> Biological organisms are best seen as representing points and peaks within evolutionary fitness landscapes. I see every such peak as a possibility imagined in the mind of God, hence possessing in Maximus's terms "logoi," divinely given patterns of being. [...] Organisms and species, then, exist in a phase space of possibilities, which the theologian will necessarily regard as ultimately deriving from God.

Here Southgate, applying the concept of the fitness landscape to Maximus' idea of creatures' divinely given *logoi*, comes very close to the argument of this book. In later work, Southgate (2018, 102) similarly further argues that "whatever in creation speaks of the work of the creator is a sign of the deep reality of God. Hence any such sign, however disturbing, or yet however apparently mundane, is a manifestation of glory." The metaphor of the "library of forms" and the dependency of evolution on the wider teleology of the cosmos, explored in this book, extend these ideas further.

But the concept of the Logos allows for linking the argument to a specifically Christian theology of nature. In contrast to approaches in which the "book of nature" and the "book of scripture" are conceived of as two wholly separate revelations, Juurikkala (2019, 347–357) thus argues that while the created *logoi* are not identified with the *Logos* (the second person of the Trinity), the *logoi* are nevertheless, particularly in the thought of

Maximus, "inseparable from him and therefore intrinsically related to all the creative, revelatory and redemptive activity of the divine Logos, Christ [...] the patristic and medieval book of nature has a substantial Christological dimension." There is room to expand here from the point of view of a theology of nature, and to formulate an understanding that makes room for biological immanent teleology as part of the broader divine purposes for the creation as a whole (Leidenhag 2019). Yet even while the development of such a wider theology of nature in this book is still incomplete, the reclaiming of the idea of design for theistic evolutionism already allows for a substantially more robust case for the compatibility of evolutionary theory with seeing the world as manifesting the wisdom of its Creator.

Notes

1. Quoted in Murray (2011, 3, 145).
2. Adam 483. Quoted in Zagzebski (2012, 116).
3. Quoted in Pearcey (1998, 84).

References

Alexander, Denis. 2018. Order and Emergence in Biological Evolution. In *The Future of Creation Order: Vol. 1, Philosophical, Scientific, and Religious Perspectives on Order and Emergence*, ed. Gerris Glas and Jeroen de Ridder, 151–170. New York, NY: Springer.

Beecher, Henry Ward. 1885. *Evolution and Religion*. New York, NY: Fords, Howard & Hulbert.

Blowers, Paul. 2016. Unfinished Creative Business: Maximus the Confessor, Evolutionary Theodicy, and Human Stewardship in Creation. In *On Earth as It Is in Heaven: Cultivating a Christian Theology of Creation*, ed. S.J. David Meconi, 174–190. Grand Rapids, MI: Eerdmans.

Cashdollar, Charles D. 1989. *The Transformation of Theology, 1830–1890: Positivism and Protestant Thought in Britain and America*. Princeton, NJ: Princeton University Press.

Clark, Kelly James, and Justin Barrett. 2011. Reidian Religious Epistemology and the Cognitive Science of Religion. *Journal of the American Academy of Religion*. 79 (3): 639–675.

Elga, Adam. 2007. Reflection and Disagreement. *Nous* 41 (3): 478–502.

Hull, David L. 1973. *Darwin and His Critics: The Reception of Darwin's Theory of Evolution by the Scientific Community*. Cambridge, MA: Harvard University Press.

Johnson, Curtis. 2015. *Darwin's Dice: The Idea of Chance in the Thought of Charles Darwin*. New York, NY and Oxford: Oxford University Press.

Juurikkala, Oskari. 2019. *The Patristic and Medieval Metaphor of the Book of Nature: Implications for Fundamental Theology*. Doctoral Dissertation, Pontificia Universita Della Santa Croce Facolta Di Teologia, Rome.

Kojonen, Erkki Vesa Rope. 2019. Is Classical Science in Conflict with Belief in Miracles? Some Bridge-Building Between Different Positions. In *God's Providence and Randomness in Nature: Scientific and Theological Perspectives*, ed. Robert John Russell and Joshua M. Moritz, 205–234. West Conshohocken, PA: Templeton Press.

Leidenhag, Mikael. 2019. Framing Christian Eschatology through Natural Teleology? Theological Possibilities and Concerns. *The Heythrop Journal*, 1–13.

Loesberg, Jonathan. 2007. Kant, Hume, Darwin, and Design: Why Intelligent Design Wasn't Science Before Darwin and Still Isn't. *The Philosophical Forum* 38 (2): 95–123.

McGhee, George. 2011. *Convergent Evolution: Limited Forms Most Beautiful*. Cambridge, MA: The MIT Press.

McGrath, Alister. 2008. *The Open Secret: A New Vision for Natural Theology*. Oxford: Blackwell Publishing.

McLeish, Tom. 2014. *Faith and Wisdom in Science*. Oxford: Oxford University Press.

Murray, Michael J. 2011. *Nature Red in Tooth and Claw: Theism and the Problem of Animal Suffering*. Oxford: Oxford University Press.

Pearcey, Nancy. 1998. You Guys Lost. In *Mere Creation: Science, Faith and Intelligent Design*, ed. William A. Dembski, 73–92. Downers Grove, IL: InterVarsity Press.

Polkinghorne, John. 2000. *Faith, Science and Understanding*. New Haven, CT: Yale University Press.

Ruse, Michael. 2018. *Darwinism as Religion: What Literature Tells Us about Evolution*. Oxford: Oxford University Press.

Shaw, Kegan. 2019. A Plea for the Theist in the Street: A Defense of Liberalism in the Epistemology of Religious Experience. *Faith and Philosophy* 36 (1): 102–128.

Simpson, William. 2021. Cosmic Hylomorphism: A Powerist Ontology of Quantum Mechanics. *European Journal for Philosophy of Science* 11: 28. https://doi.org/10.1007/s13194-020-00342-5.

Southgate, Christopher. 2008. *The Groaning of Creation: God, Evolution, and the Problem of Evil*. Louisville, KY: Westminster John Knox Press.

———. 2018. *Theology in a Suffering World: Glory and Longing*. Cambridge: Cambridge University Press.

Stanley, Matthew. 2015. *Huxley's Church and Maxwell's Demon: From Theistic Science to Naturalistic Science*. Chicago, IL: The University of Chicago Press.

Wahlberg, Mats. 2012. *Reshaping Natural Theology: Seeing Nature as Creation*. London: Palgrave Macmillan.

Zagzebski, Linda Trinkaus. 2012. *Epistemic Authority: A Theory of Trust, Authority and Autonomy in Belief*. Oxford: Oxford University Press.

Index[1]

A
Abraham, William, 25, 31, 45n6, 81
Aguilar-Rodriguez, José, 121
Ahnert, S.E., 122
Alexander, Denis, 123, 131, 134n13, 209
Alston, William, 19, 34
Analogical argument, 75
Analogy, 1, 2, 23–25, 69, 71, 75, 76, 84, 88, 100, 102–104, 109, 129–131, 166, 167, 170–172
Anthropomorphism, 8, 22, 79, 80
A priori, 23, 59, 86, 149, 207
Aquinas, 25, 27, 46n17, 71–74, 76
Aquino, Frederick D., 31
Architect analogy, 100–103, 134n3, 167, 170–172
Ardern, Zachary, 127, 132
Argyll, George, 35, 38, 157
Ariew, André, 74, 104, 196n2
Aristotelian, 73, 74
Aristotle, 18, 72
Artigas, Mariano, 133n1
Augustinus, Aurelius, 27, 181
Automated factory thought experiment, 166–168, 185, 206
Axe, Douglas, 119, 120, 134n12, 196n6
Ayala, Francisco, 3, 97, 175, 177, 178, 197n10

B
Baker-Hytch, Max, 35, 43, 197n11, 197n14
Banner, Michael C., 77
Barbour, Ian G., 15, 20, 29
Barnes, Luke, 65, 66, 79
Barrett, Justin, 39, 40, 42, 207
Barrow, John, 66
Barth, Karl, 18
Barton, R., 149

[1] Note: Page numbers followed by 'n' refer to notes.

E. V. R. Kojonen, *The Compatibility of Evolution and Design*, Palgrave Frontiers in Philosophy of Religion, https://doi.org/10.1007/978-3-030-69683-2

Beecher, Henry Ward, 205, 206, 208
Behe, Michael J., 70, 73, 89n6, 115–118, 122, 147, 158, 179
Benton, Matthew A., 43
Berhow, Michael, 110, 197n10, 197n13
Berlinski, David, 105, 107
Bertka, Constance M., 127
Best, Robert B., 119
Bhaskar, Roy, 149
Bignon, Guillaume, 197n19
Biocentric, 186
Bird, Alexander, 78
Blancke, Stefaan, 112
Bloom, Paul, 39
Blowers, Paul, 212
Boethius, Ancius Manilius Severus, 178
Bonhoeffer, Dietrich, 10n2
Boudry, Maarten, 71, 112, 116
Bowler, Peter, 98, 135n14
Braeckman, Johan, 116
Brandon, C., 134n10
Brooke, John Hedley, 18, 19, 45n1, 155
Brown, C. Mackenzie, 21
Brown, James Robert, 82
Brown, William P., 10n1
Browne, Janet, 11n4, 100
Burns, Aaron, 36
Burrell, David B., 25

C

Calvin, Jean, 197n19
Capitalism, 104
Carroll, Sean, 123
Causal responsibility, 156, 169, 180, 190, 211
Cavanaugh, William T., 29
Chan, Kin, 46n18
Chance, 1, 2, 26, 28, 32, 69, 70, 72, 73, 78, 89n6, 97, 98, 102, 103, 105, 106, 108–111, 123, 127, 134n4, 145, 153, 154, 162–164, 168–170, 176, 179, 188, 197n13, 206
and necessity, 28, 97, 109–111, 127, 164, 179
Chignell, Andrew, 34
Clark, Kelly James, 39, 40, 207
Classical theism, 24, 25
Cleland, Carol, 62
Coakley, Sarah, 176
Cognitive science of religion, 39, 42
Collins, Robin, 63, 65–68, 78, 89n3, 173
Common sense, 82, 164, 173, 196n6, 206, 207
Convergence, 21, 77, 124–128, 131, 132, 168
Conway Morris, Simon, 125–128, 168, 172
Copan, Paul, 25
Corabi, Joseph, 197n13
Cosmological argument, 64, 160
Coyne, Jerry, 45n2, 107
Craig, William Lane, 18, 25, 31
Craver, Karl, 11n3
Creation out of nothing, 25, 180, 181, 191, 193, 197n18, 197n20
Crick, Francis, 38
Crop circle thought experiment, 150, 159
Culture, 194
Cunningham, Conor, 3, 13, 14, 22, 183

D

Darwin, Charles, 2, 4, 7, 10, 18, 26, 30–33, 35, 38, 43, 59, 60, 87, 89n4, 97–105, 107, 109, 117, 134n3, 135n14, 145, 153, 157, 167, 170, 172, 176, 177, 188, 206, 208

Davies, Brian, 197n19
Dawes, Gregory, 61, 64, 110
Dawkins, Richard, 14, 30, 32, 38, 47n20, 59, 60, 106–108, 110–112, 114, 115, 134n4, 145, 207
De Cruz, Helen, 21, 39–41, 65, 81, 192
De Duve, Christian, 98
De Ridder, Jeroen, 45n3
De Smedt, Johan, 21, 39–41, 65, 81, 192
Death, 29, 146, 149–151, 177, 182, 183, 186
Debunking argument, 40, 43, 44
Defeater, 35, 40, 43, 45, 88, 157–159, 163, 164, 196n4
Dembski, William A., 2, 3, 23, 30, 69, 84, 89n6, 109–116, 118, 120, 134n8, 147, 179, 185, 197n13, 197n17
Dennett, Daniel, 103, 111, 130, 158, 164, 172
Denton, Michael, 69, 125, 129, 134n10
Design discourse, 8, 9, 34, 35, 40, 146, 157, 159–162, 209
Determinism, 89n4, 97, 174, 197n19, 211
Dickins, T., 149
Directionality, 5, 9, 72, 128, 129, 132, 155, 171, 206, 209–211
Divine action, 29, 44, 47n19, 64, 185, 187, 188, 196n3
Dobzhansky, Theodosius, 7
Doran, Chris, 197n10
Dougherty, Trent, 36, 46n7, 177, 182, 197n19
Doyle, Arthur Conan, 78
Draper, Paul, 116, 118, 175, 181

E
Economics, 104, 109
Elga, Adam, 207
Ellis, George F., 28
Elsberry, Wesley, 80
Engineering, 130, 173, 179, 195
Epistemic fit, 18
Epistemology, 15, 34–36, 65, 90n11, 151
Erwin, Douglas H., 128, 131
Evans, C. Stephen, 31, 32, 34, 85
Evidence, 30, 36, 85, 114, 118, 128, 132, 146, 164, 175, 176
Evil God hypothesis, 197n11
Evolution
of the eye, 2, 31, 105, 163, 175
as a search engine, 67, 113, 123, 126, 131
Ewert, Winston, 112, 114
Exaptation, 119
Extended synthesis, 99, 132, 210

F
Faith and reason, 17, 18, 30, 35, 36, 40, 74, 81, 85, 100, 195
Fehige, Yiftach, 82
Fergusson, David, 28, 177
Feser, Edward, 45n2, 60, 61, 72–74, 85, 90n12, 150
Fideism, 18, 30
Fine-tuning, 32, 63, 65–68, 83, 87, 119, 122, 123, 126, 132, 133, 145, 156, 162, 164, 167, 168, 173, 174, 206, 209
Firing squad analogy, 67
Fodor, Jerry, 99, 133n2
Fortuna, Miguel A., 121
Freeman, Scott, 98
Free will, 97, 174, 187–190

Fry, Iris, 127
Fugue, 169, 170
Futuyma, Douglas, 26

G
Ganeri, Martin, 21
Garvey, Jon, 22, 176
Gauger, Ann, 119
Genetic algorithm, 108, 109, 111, 112
George, Marie, 46n14, 73, 74, 76, 163, 164
Geraint, Lewis, 65
Gettier, Edmund, 33
Giberson, Karl, 28, 133n1
Gishlick, Alan, 117
Glass, David, 85, 131, 148, 151, 154, 155
Glory, 1, 19, 25, 180, 186, 189, 193, 198n21, 212
Godfrey-Smith, Peter, 71
God of the gaps, 3, 10n2, 13–17, 44, 60, 86, 165
Gonzales, Guillermo, 66
Goodwin, Brian, 125, 130
Gordenin, Dmitry A., 46n18
Gould, Stephen Jay, 70, 119, 124, 127, 128, 167, 168, 170
Gray, Asa, 4, 5, 88, 98, 100, 102, 104, 135n14, 154, 167, 206
Gremlin thought experiment, 80, 81
Griffiths, P. E., 41, 129
Groothuis, Douglas, 45n4
Guided mutations, 28, 29, 100–102, 108, 112, 115, 148, 153, 155, 156, 196n3

H
Häggström, Olle, 113, 114, 134n8
Halvorson, Hans, 66
Hanby, Michael, 3, 4, 23, 24
Harrison, Peter, 45n1
Harrison, Victoria, 197n9
Haught, John F., 15, 26, 197n15
Henrich, Joseph, 189
Henriksen, Jan-Olav, 40
Herron, John C., 98
Hewlett, Martinez, 26, 182
Hietanen, Johan, 45n3
Himma. Kenneth, 110
Hoggard Creegan, Nicole, 132
Holder, Rodney, 20
Hossenfelder, Sabine, 82
Hoyle, Fred, 134n7
Hull, David, 208
Hume, David, 16, 18, 38, 59, 60, 64, 68, 75, 76, 80, 86

I
Ichneumonidae, 175
Inference
  to the best explanation, 7, 75–79, 81, 88, 106
  to the only explanation, 78, 79
Infinite monkeys thought experiment, 103
Information, 21, 70, 71, 84, 111–115, 148, 161
Intelligent Design, 2, 7, 8, 13, 26, 61, 77, 116, 148, 208
Inter-religious dialogue, 21
Intervention, 28, 109, 122, 146, 184, 185, 192
Intuition, 5, 38, 98, 133, 134n12, 157, 162, 206, 207
Irreducible complexity, 115, 116, 118–122, 158

J
Jablonka, Eva, 46n18
Jiménez, Jose, 124
John, J.D., 108
Johnson, Curtis, 11n4, 100, 206

Johnson, Phillip E., 22, 23
Jong, Jonathan, 40, 42
Juthe, A., 76
Juurikkala, Oskari, 19, 212

K
Kacar, Betül, 131
Kahane, Guy, 40
Kalam, 64
Kant, Immanuel, 18, 37, 38, 64, 208
Kauffman, Stuart A., 111, 128
Kelemen, Deborah, 39, 42
Kim, Jaegwon, 148, 196n1
Kingsley, Charles, 5, 205, 208
Kitcher, Phillip, 90n8
Koistinen, Timo, 45n5
Kojonen, Erkki Vesa Rope, 2–4, 8, 11n2, 15, 16, 18, 23, 30, 46n13, 61, 70, 71, 77, 81–83, 89n5, 89n6, 90n7, 114–117, 147, 148, 158, 159, 162, 165, 166, 179, 185, 193, 196n7
Koonin, Eugene, 125, 128
Koons, Robert C., 63
Koperski, Jeffrey, 62, 75, 77, 150
Kretzmann, Norman, 27

L
Laland, Kevin, 99, 149
Lamb, Marion J., 46n18
Lamont, John, 21, 160
Lane Ritchie, Sarah, 46n13
Launonen, Lari, 39
Laws of form, 9, 123, 125, 126, 129, 130, 132, 133, 145, 152, 154, 168, 173, 194
Leech, David, 40
Leibniz, Gottfried, 184, 185
Leidenhag, Mikael, 70, 196n1, 213
Leisola, Matti, 26, 46n16
Lenoir, Timothy, 38
Lenski, Richard, 108, 112
Leslie, John, 67, 83
Level-shifting, 9, 147, 150, 165
Levering, Matthew, 19, 25, 27
Lewens, Tim, 99
Lewontin, Richard, 70, 170
Library of forms, 121–123, 126, 127, 129–131, 133, 152, 191, 192, 209
Lipton, Peter, 77, 78
List, Christopher, 174
Livingstone, David N., 109
Loesberg, Jonathan, 47n20, 60, 208
Loikkanen, Juuso, 89n6, 110, 178, 197n12
Loke, Andrew Ter Ern, 18, 31, 64
Louis, Ard, 126
Love, 18, 24, 25, 29, 97, 187, 188
Luther, Martin, 27, 178, 197n19
Lyell, Charles, 100

M
MacIntyre, Alasdair, 20
Macready, William, 111
MacSwain, Robert, 22
Malaria, 179, 180
Marks, Robert J. II, 111, 114
Marshall, Craig, 125
Mascord; Keith A., 35, 45n5
Matzke, Nicholas J., 117, 118
Maximus the Confessor, 212
Mayr, Ernst, 149, 196n2
McBrayer, Justin P., 177, 197n19
McCall, Thomas, 29
McCartney, Mark, 131, 155
McCord Adams, Marilyn, 25
McCosh, James, 155
McDowell, John, 47n22
McGhee, George, 77, 125–129, 131, 134n11, 210

McGlynn, Aidan, 90n11
McGrath, Alister, 3, 6, 10n1, 18, 20–22, 35, 46n9, 47n23, 62, 77, 89, 149, 176, 178, 181, 207
McGrew, Lydia, 63, 76
McGrew, Timothy, 66, 79
McLeish, Tom, 20, 124, 177, 192, 211
McMullin, Ernan, 27, 181
Meek, Esther, 45n7
Meester, Roy, 112
Message thought experiment, 83, 84, 165
Metaphysics, 16, 26, 60, 64, 65, 67, 73, 74, 87, 90n9, 114, 123, 133, 181, 192, 207, 208
Methodological naturalism, 23, 86
Meyer, Stephen C., 71, 78, 120, 148
Miller, Calum, 176
Miller, Kenneth, 117
Miracle, 127, 173, 184
Moon, Andrew, 35
Moral responsibility, 156, 180, 187, 189, 190, 194, 195, 197n19
Moreland, J. P., 18, 23, 31, 114
Moritz, Joshua M., 135n15, 187
Morphospace, 126, 129
Moser, Paul K., 45n7
Mullen, Thomas, 34, 37
Müller, Gerd B., 99
Multiverse, 67, 68, 89n3, 128
Murphy, Nancey, 33
Murray, Michael J., 41, 177, 185, 186, 197n16, 197n18, 213n1
Musgrave, Ian, 117, 118

N

Nagel, Thomas, 163
Narveson, Jan, 80
Natural
- evil, 185
- revelation, 20, 46n8
- selection, 2, 26–28, 43, 89n4, 98–104, 106–109, 112, 117, 118, 120–123, 128, 129, 132, 148, 152, 163, 167, 174, 181, 197n14
- theology, 3, 17–22, 26, 31–33, 45n4, 45n5, 45n7, 46n8, 46n9, 46n10, 63, 68, 86, 87, 154, 162, 174, 178, 179, 194, 195, 196n4, 197n9, 206, 210

Naturalism, 16, 30, 68, 86, 179, 207, 208, 210
- of the gaps, 16, 60, 69, 86, 87, 103, 207

Neptune, Clinton, 46n7
Neutral evolution, 70, 134n5
Newman, John Henry, 21
Newton, Isaac, 176, 184
Newton, William, 72, 74
Niche construction, 99, 189, 210
Nicholson, Daniel J., 71
Nilsson, Dan-e, 107
Noble, Denis, 71, 106
Nowak, Michael, 176

O

Oliver, Simon, 74
Oord, Thomas, 188
Oppy, Graham, 76
Origin of life, 60, 127, 128, 132, 161
Orr, Allen, 60, 117, 118
Oslington, Paul, 104
Owen, Richard, 123

P

Page, Ben, 197n11
Pak, Kenneth, 198n20
Paley, William A., 1–3, 10n1, 13, 23–25, 35, 38, 43, 46n14,

46n17, 59, 60, 69–76, 82, 88, 98, 102, 103, 105, 106, 134n3, 147, 149, 157, 163, 176, 179
Pallen, Mark J., 117, 118
Palmerino, Carla Rita, 82
Pannenberg, Wolfhart, 46n8
Pascal, Blaise, 31, 85
Pearcey, Nancy, 208, 213n3
Peckham, John, 182, 187
Peels, Rik, 36, 45n3, 82, 175, 183, 197n14, 197n17
Peirce, C. S., 77
Pelger, Susanne, 107
Pence, Charles H., 46n18
Pennock, Robert, 84, 90n8
Perakh, Mark, 110, 111
Perlman, Mark, 70
Perry, John, 46n13
Persson, Johannes, 151
Peters, Ted, 26, 182
Philipse, Herman, 47n19, 61
Phillips, D. Z., 45n5
Philo, 38, 47n21
Philosophy of explanation, 61, 63, 64, 66–68, 70, 74, 77–80, 82, 86, 88, 90n8, 110, 148–152, 154, 156, 166, 167, 193, 194, 196n4, 208, 209
Piattello-Palmarini, Massimo, 99, 133n2
Pigliucci, Massimo, 71, 99, 129, 134n11
Plantinga, Alvin, 8, 9, 26, 34–36, 45n5, 60, 157–161, 187, 196n5
Platonic, 74, 125, 129, 152
Polkinghorne, John, 197n15, 209
Pollock, John L., 33
Preadaptation, 119, 126
Pressman, Abe D., 125
Primary cause, 27, 100, 102, 150
Problem
of evil, 29, 156, 175, 178, 179, 186, 195, 212
of good, 177, 195, 197n11
of underdetermination, 79
Process theism, 29
Protein, 122
evolution, 119–121, 124–126, 132, 168
Proximate explanation, 153
Pruss, Alex, 31
Pust, Joel, 90n10

R

Ramsey, Grant, 46n18
Ramsey, William, 22
Rasmussen, Joshua L., 31
Ratzsch, Del, 21, 34, 37, 38, 62, 75, 77, 150, 164–168, 206
Ray, T., 108
Re Manning, Russell, 18
Reading in *vs.* reading out of, 5, 21, 22, 160
Reductio ad absurdum, 22, 130, 173
Reformed epistemology, 35
Reid, Thomas, 34
Reiss, Michael, 89n4
Revelatory potential, 5, 14, 30, 46n8, 146, 162, 188, 194, 205, 213
Richards, Jay W., 66
Rickabaught, Brandon, 46n7
Ridley, Mark, 98
Riskin, Jessica, 38, 109
Rockoff, Hugh, 104
Rosenberg, Alex, 15, 47n19
Rothuizen-van der Steen, J., 47n19
Ruse, Michael, 2, 3, 30, 31, 59, 70, 124, 197n13, 205, 207
Russell, Robert J., 29, 180

S

Saarinen, Henrik, 45n3
Sagan, Carl, 84
Salmon, Wesley, 62

Schaffer, Jonathan, 168
Schmidt, William E., 150
Schneider, John R., 177, 182
Schupbach, Jonah, 76, 148, 152, 154
Scientism, 8, 15–17, 22, 30, 43, 44, 45n2, 45n3, 61, 64
Secondary cause, 4, 100
Sedley, David, 21, 134n4, 146
Sehon, Scott, 89n2
Sennett, 45n4
Sepkoski, David, 99
SETI, 84
Sexual selection, 99, 189
Shakespeare, William, 107, 110
Shanks, Niall, 104
Shaw, Kegan, 32, 206
Shostak, Seth, 84
Silva, Ignacio, 46n15
Simon, Dan, 108
Skogholt, Christoffer, 196n1
Smith, Adam, 104
Smith, James K. A., 29
Snellman, Lauri, 183
Sober, Elliott, 28, 59, 80, 81, 85, 98
Sollereder, Bethany, 29, 46n15, 187–189, 197n17
Southgate, Christopher, 29, 47n23, 174, 177, 178, 181, 187, 189, 197n15, 212
Specified complexity, 89n6, 110, 111, 114
Stanley, Matthew, 206
Stenger, Victor, 66
Stenmark, Mikael, 18, 45n3
Sterba, James P., 186
Stitch, Stephen P., 36
Structuralism, 125, 129, 130, 132, 172
Stump, Eleonore, 25, 177
Sudduth, Michael, 45n4
Susskind, Leonard, 67
Swanke, Johannes, 27
Swinburne, Richard, 15, 19, 36, 46n10, 61, 63, 64, 79, 89n1, 178

T

Talbert, Matthew, 190
Taliaferro, Charles, 64
Tawfik, Dan S., 120
Taylor, S.V., 119
Technology, 69, 172–174, 191, 194
Teleology, 3, 5, 6, 9, 17, 31, 69–76, 83, 84, 86–89, 89n4, 98, 114, 119, 122, 132, 133, 134n14, 145, 146, 148, 149, 152–154, 156–162, 165, 166, 169, 173, 174, 177, 180, 189, 191, 194, 195, 205, 207–209, 211–213
Tertullianus, Quintus Septimius Florens, 42
Theist on the street, 32, 156, 162, 164, 206
Theodicy, 68, 156, 175, 177–180, 184, 186–189, 191–193, 195, 197n10, 197n15, 198n21
Theological language, 25
Theology of nature, 5, 20, 22, 44, 81, 159, 162, 167, 194, 195, 213
Thomism, 133
Thought experiment, 80, 81, 103
Tian, Pengfei, 119
Timpe, Kevin, 189
Tipler, Frank, 66
Tokuriki, Nobuhiko, 120
Tomasello, Michael, 189
Tooze, J., 134n10
Tornado in a junkyard, 134n7
Torrance, Alan, 29, 46n13
Tracy, Thomas, 29, 153
Turner, Denys, 25, 74

U
Ulanowitz, Robert, 28

V
Vainio, Olli-Pekka, 18, 24, 31, 42, 45n3, 197n19
van Eyghen, Hans, 41
van Huyssteen, Wentzel, 21
van Woudenberg, René, 45n3, 47n19, 82
Vatican International Theological Commission, 27
Venkataraylu, N.V., 108
Veridical illusion, 33, 43
Vestrup, Eric, 66
Visala, Aku, 39–42, 190, 196n2, 197n19
Vodder, Hans, 196n7
von Balthasar, Hans Urs, 23
von Wachter, Daniel, 36
Vrba, Elisabeth, 119

W
Wagner, Andreas, 46n18, 119–122, 125, 127, 129–131, 134n12, 152
Wahlberg, Mats, 5, 33, 34, 37, 46n11, 47n22, 160, 161, 163, 169–171, 183, 186, 189, 196n8, 197n15, 207
Walsh, Denis, 70
Ward, Keith, 131
Watchmaker analogy, 2, 23, 69, 75, 83, 106, 115, 122, 147
Weinrich, Daniel, 124
Whewell, William, 34
White, Thomas Joseph, 45n4
Wichmann, Stefan, 127
Wigner, Eugene, 173
Wiker, Benjamin, 162, 173
Wilkins, J.S., 40, 41, 80
Williams, Thomas, 25
Williamson, Timothy, 90n11
Willmer, Patricia, 125
Witt, Jonathan, 162, 173
Wolf, Yuri A., 125
Wolpert, David, 111, 134n8
Woodward, James, 61, 62

Y
Ylikoski, Petri, 82, 151

Z
Zagzebski, Linda, 213n2

CPSIA information can be obtained
at www.ICGtesting.com
Printed in the USA
LVHW011259140821
695295LV00004B/747

9 783030 696825